APOCALYPSE SOON?

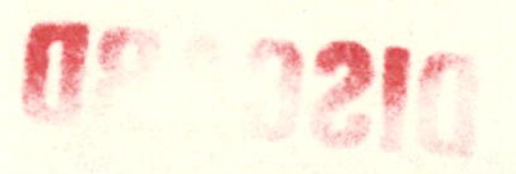

Apocalypse Soon? Wagering on Warnings of Global Catastrophe

Stephen F. Haller

McGill-Queen's University Press
Montreal & Kingston · London · Ithaca

ISBN 0-7735-2437-1 (cloth)
ISBN 0-7735-2438-x (paper)

Legal deposit fourth quarter 2002
Bibliothèque nationale du Québec

Printed in Canada on acid-free paper that is 100% ancient forest free (100% post-consumer recycled), processed chlorine free.

This book has been published with the help of a grant from the Humanities and Social Sciences Federation of Canada, using funds provided by the Social Sciences and Humanities Research Council of Canada.

McGill-Queen's University Press acknowledges the support of the Canada Council for the Arts for our publishing program. We also acknowledge the financial support of the Government of Canada through the Book Publishing Industry Development Program (BPIDP) for our publishing activities.

National Library of Canada Cataloguing in Publication

Haller, Stephen F. (Stephen Francis), 1958-
Apocalypse soon?: wagering on warnings of global catastrophe / Stephen F. Haller.
Includes bibliographical references and index.
ISBN 0-7735-2437-1 (bound) – ISBN 0-7735-2438-x (pbk.)
1. Disasters – Forecasting. 2. Disasters – Forecasting – Mathematical models. I. Title.
H61.25.H34 2002 363.34'01'12 C2002-901633-9

This book was typeset by Dynagram Inc. in 10/13 Sabon.

For Karen Hennig
Diogenes Can Stop Searching

Contents

Acknowledgments

Inspired by Somerset Maugham's *The Razor's Edge*, I spent many months living in a small village in Bali, Indonesia – to "loaf" as his protagonist would call it (meaning, of course, to think about serious philosophical issues). It was here that I first started thinking about risk, and eventually decided to write about environmental risk. Living on a paradise island made me realize all the more how important was my topic. Bali is facing the acute stresses of overpopulation and pollution, and while walking along beaches in the equatorial South Pacific under star-draped skies, or scuba diving on the complicated living systems of coral reefs of exquisite beauty, it is hard not to experience the sublime and recognize the value of what is at risk. It was this experience, along with the knowledge of the fragility of systems like coral reefs, that focused my thinking on Pascalian wagers and environmental risk. I would like to thank all the people who made my stay there so pleasant.

I have many more people to thank for their help over the years. Foremost among these is Mark Mercer. Mark is the best philosopher I have ever met, and I have benefited greatly from his insight. He has helped me sort out my most vexing philosophical confusions, and offered encouragement when it was most needed. I could not have completed this project without his help. Bill Duggan has been my lifelong friend and his unflagging support has been invaluable. Special thanks is extended to Karen Wendling, whose insightful comments helped to guide the direction of my research through the maze of possibilities. Discussions with Steven Robinson helped me to focus the questions I needed to answer. I would like to thank Catherine Beattie, Elizabeth Boetzkes, William Hughes, Hugh Lehman, Bernard Lightman, Denis Lynn, David Martens, and Paul Thompson for their suggestions on early drafts.

Two persons deserve special mention. John Leslie's skeptical questioning kept me focused on the details and Michael Ruse helped me to see the big picture. Any mistakes in reasoning remain my own, of course. The drawings and graphs are supplied by Andrew Montgomery; many thanks. I would like to thank my mother, Eileen Haller, for inspiring me to higher education, and my father Frank Haller for his support along the way. I take this opportunity to also express thanks to Gord Ross, Adam Dodd, Jim Gerrie, and my brothers and sisters.

Preface

In a small corner of the South Pacific, on the island of Bali, seven or eight women are ankle-deep in a shallow, muddy pond. They chant rhythmically as they plant the young shoots of rice. Nearby, the men are mending the mud walls that curve sinuously around the contours of the mountain and that control the water flow to hundreds and hundreds of such little ponds. As one of them stands up to stretch his back, his gaze lingers on the scene – a thousand shades of green all the way down to the sea – and he thinks that his life is ephemeral compared with the permanence of the surroundings. *These rice paddies have been here for centuries and centuries, and will continue to look pretty much the same after I am gone*, he thinks. *The important thing is that life go on. I am participating in the stream of life and my children will soon be here with their backs bent over these fields.*

It is hard to imagine how such seemingly innocuous activities – growing rice and having children – could threaten the renewal of this idyllic scene, and yet, that is what might be happening. The population of Bali is already too large to be accommodated on that little island, and hundreds of thousands of people are being transplanted to more remote regions of the Indonesian archipelago. This human dispersal is creating as many problems as it solves, political unrest and the destruction of wildlife habitat chief among them. Growing rice to feed this burgeoning population is releasing ever larger amounts of methane, a gas that some people claim is a major contributor to the greenhouse effect that might ultimately change the climate of the entire Earth in ways that could be catastrophic.

Around the world, billions of other people are also getting up and going to work, heating their homes, and having children. These daily life activities might also be contributing to a catastrophe that will change Earth's ability to support this large population.

As someone with environmentalist values, I was interested in looking into claims of potential global crises. I started out with the not uncommon view that Earth is in peril because of a mixture of human carelessness, high technology, and greed. We might be destroying Earth merely to "expedite the affairs of our pot and kettle,"to borrow a phrase from Ralph Waldo Emerson's essay on *Nature*.[1] I set out to defend the idea of precaution. I, myself, do not mind the monkish life, and I thought if everyone else could give up what Pascal calls "noxious pleasures, glory and good living," then the world would be safe from catastrophe.[2] The more I examined the claims that doomsday is approaching, the less convinced I became that anybody knows whether we are on the brink or not. Our best science can give us no assurance that doomsday is either likely or unlikely. Furthermore, should catastrophe befall us, its cause might well be ordinary, innocent living, and not greed, carelessness, or high technology.

In this book, I will make judgments about which predictions of catastrophe are well supported by the evidence, and which are rendered improbable by the evidence. How we are to evaluate predictions of catastrophe poses many interesting questions in epistemology and the philosophy of science. I believe I have reconstructed the best arguments available. However, you are free to disagree and question my experts. I am not committed either way toward any particular global model. My thesis will not stand or fall if the nuclear winter hypothesis is reinstated, nor if we find that ozone depletion is not the result of CFCs, because my thesis is not about attacking or defending any particular prediction of potential catastrophe. My concern is about the general problem of what to do in cases where we are asked to take action meant to avoid catastrophe before we have compelling evidence of the likelihood of the catastrophe, and I use nuclear winter, ozone depletion, and global warming as examples only. My interest is not in evaluating particular doomsday scenarios, but in making explicit and defending the criteria that should be used in evaluating them. The central issue I deal with is what constitutes the most rational approach to the possibility of catastrophe, when we have no reliable estimates of the likelihood of catastrophe. You will find that, in the end, I still side with the sentiments of Pascal and defend precaution as the rational approach to take when we believe catastrophe is possible but have no idea about how likely it is.

APOCALYPSE SOON?

Introduction

Will the activity of humans soon render Earth unable to support the life that it now supports? More than a few scientists have claimed to have shown that, should present habits of production and consumption continue, we will soon reach the limits of Earth's capacity to support its inhabitants. Population, food and resource consumption, and industrial emissions have been increasing at exponential rates, and the speed of the increase has resulted in environmental and social problems on a scale far surpassing anything we have encountered previously. Potential crises are global in scope, and the possible consequences are catastrophic. We are risking the very capacity of Earth to support any life, or at least our own capacity for survival as a species. In *It's a Matter of Survival*, Anita Gordon and David Suzuki claim "we are the last generation on earth that can save the planet."[1] If this claim is correct, any humans lucky enough to survive the impending juggernauts of, say, overpopulation, global warming, nuclear winter, or ozone-layer depletion, will, it is asserted, find themselves in a world irreversibly damaged – a sun-baked, tornado-swept, and life-impoverished planet.

Such warnings are usually accompanied by an urgent demand for action to avert the potential catastrophe. However, these warnings are a mixed blessing. On the one hand, they might well spur us to action that will successfully prevent a disaster. On the other hand, warnings of apocalypse might paralyze us into a state of indecision, or lead to an overreaction that squanders limited resources in attempts to avert a catastrophe that is not, in fact, forthcoming. Jonathan Swift, in his description of Laputa in *Gulliver's Travels*, poked fun at those who are in a perpetual state of anxiety over potential doom. The cloud-dwelling Laputians were incessant doomsayers – inventing theories about how the sun might become encrusted in its own effluvia and, thus, cease to warm and light Earth. The Laputians would greet each other each morning with worries about the possibility of destruction and were continually estimating the odds of a comet crashing into Earth, or of Earth's orbit decaying soon and our planet crashing into the sun. However, this activity was not obviously advantageous to them. Given the uncertainty of the predictions, the prophecies of doom places the Laputians "under continual disquietudes, never enjoying a minute's peace of mind ... They are so perpetually alarmed with the apprehensions of these and the like impending dangers, that they can neither sleep quietly in their beds, nor have any relish for the common pleasures or amusements of life."[2]

We are in much the same position today. We also have predictions of a catastrophic end to our environment and ways of life and are greeted daily in our magazines and newspapers with discussions of potential environmental disaster. Swift suggests that the Laputians wasted their energies worrying about threats that were not well enough founded to justify their reaction to them. We do not want to make the same mistake as the Laputians and worry unnecessarily about things that may not come true, nor do we want to take drastic action and make sacrifices today for the sake of avoiding a phantom future. Faced with modern apocalyptic visions, we risk being paralyzed with fear like the Laputians because we do not know how to assess the accuracy of these visions. Some potential catastrophes are beyond our ability to address, and, therefore, a rational person should not worry about how to address them. However, there might be some that we could and, thus, perhaps would like to address. How can we tell the difference? How can we tell whether current fears of catastrophes are any less unfounded than those satirized by Swift? Is it rational to invest in preventive measures before we have ascertained whether they are not just an overreaction to irrational fears? We have heard doomsday predictions too many times before, so we need good reasons to act on any newcomers. Yet, some global hazards might, in their very nature, be such that they cannot be prevented unless pre-emptive action is taken immediately – that is, before we have evidence sufficient to convince ourselves of the reality of the threat. Unless we act now on uncertain claims, catastrophic and irreversible results might unfold beyond human control. The longer we wait, the more difficult it might become to intervene.

If we are to understand whether claims of potential catastrophe are credible, and also whether it is possible to intervene, we need an understanding of the mechanisms by which the catastrophes would befall us. We create models to supply us with this understanding. A good model is an idealized representation of some system, a representation able to reveal general patterns in that system's behaviour and to provide predictions of what will happen when certain changes occur in the system. Thus, for example, one well-constructed global circulation model of Earth's climate predicts that the global average temperature will increase with increasing concentrations of carbon dioxide. Furthermore, by identifying where a model is stable and where it is sensitive to change, modellers hope to gain information useful in deciding how best to influence real-world systems in order to prevent or minimize predicted catastrophes.

Unfortunately, we have little reason to place much confidence in the ability of current models of global systems to make reliable predictions. The reliability of the predictions generated by these models is questionable because, unlike models of smaller systems, models of global systems cannot be tested against their predictions and improved by trial and error. At the same time, the stakes are high. Earth might be at risk, and, perhaps, something should be done – but the advice of those who interpret the results of the models is usually very costly. To prevent global warming, we are told, we must change our transportation and agricultural practices. To prevent economic or agricultural collapses, we must find ways of reducing population growth. Given the high costs of precaution, political bodies are forced to make a decision regarding the rationality of acting upon the assumption that the model is providing a realistic representation. To address the problem of global catastrophic risks, we need to know both what these risks are and what we are willing to do about them. Thus, the book is divided into two parts – what to believe and what to do.

In part 1, I evaluate claims about the risk of global catastrophe should certain current trends continue. What is a model? What criteria are we to use to determine which models are merely fanciful speculation and which are sound and worth considering in our practical deliberations? I examine the idea that the ability to supply intellectually satisfying explanations (to disinterested and well-informed experts) might be a sufficient criterion of a model's worth. This discussion is necessary because the usual method of assessing a model by measuring its predictive success is not available for models of the kind that predict future global catastrophes like nuclear winter or global warming. Obviously, observations that a model's predictions of nuclear winter or global warming have been borne out in the course of time would come too late to be useful. In the absence of empirically demonstrable predictive power, are reasonable analyses of societal, economic, and environmental mechanisms that portray the possibility of a disastrous future sufficiently convincing? To answer this, I present in chapter 1 four examples of global systems models that predict catastrophe (given certain initial conditions) and explain their goals as well as give some of their history. Next, in chapter 2, I discuss the nature of models in general and show how their two goals of prediction and explanation can be at odds with each other. Modellers must often choose between predictive power and the complicated details needed for explanation. This implies that models must be understood as being designed for special purposes, rather than as definitive representations of reality. In chapters 3–6, I draw on arguments concerning both the conditions of a good prediction and the predictive weakness of some ecological models, to conclude that the models of global systems that are predicting various catastrophes leave us uncertain as to the likelihood of these catastrophes. The global models I investigate can show what is a possible future, but cannot provide reliable estimates of probability for that future.

Part 2 examines the question of decision-making and action in the face of uncertainty. Acting precautiously will involve sacrifices and losses of opportunity, and one wonders about the rationality of adopting a precautionary strategy in every case. If it is rational to invest in measures that will ameliorate the chances of global warming, would it also be rational to invest in research, say, to learn how to avoid impact from intergalactic debris? (One estimate of the "comparatively trivial expenditures" needed for detecting meteorite threats is approximately US$10 million a year.[3]) Concern about a potential catastrophe necessitates making some difficult decisions with respect to trade-offs between safety and opportunity, as well as between individual freedoms and the overall benefits for society. Choosing to reduce emissions of greenhouse gases might stifle economies and restrict the mobility of individuals. Choosing to preserve large areas of wilderness will limit opportunities for development, as well as require a huge investment of effort and money.

Decision theory is, in many situations, a useful tool for guiding actions. After assigning estimates of value and probability to the various possible outcomes of our actions, we could pursue the course of action that we believe to have the highest expected utility. However, in situations of uncertainty where the possible consequences of one course of action (or failure to take action) are catastrophic, even estimates of very low probability for this outcome will be worrisome. For example, a 1999 "fly-by" of the NASA *Cassini* spacecraft was controversial because there was a small, but potentially catastrophic, risk that the spacecraft could crash into Earth's atmosphere, dispersing its plutonium fuel across a large portion of the globe. Representatives from NASA claimed that the risk of such a disaster was negligible, estimating the chances to be about 1 in 1.2 million.[4] There are many such risks where the possible consequences are catastrophic, yet the estimated probability of occurrence is very low or, perhaps, even approaching zero. If policy-makers decide to be cautious whenever faced with one of these dilemmas, the costs in resources and lost opportunities will be very high. Deciding between the two risks (a low risk of catastrophe or a high risk of wasteful precaution) is controversial. On the one hand, there are those who argue that we should not be afraid of the unknown. They argue that policy-makers should not react to scenarios of catastrophe until the evidence for them is strong. On the other hand, there are those who advocate maximum precaution. A policy of risk-taking might, sooner or later, result in catastrophe. The precautionary principle, which has been included in many international environmental agreements, advises that lack of scientific evidence for a claim should not be taken as a reason for exercising a lack of caution when the risk is high. When the risk is high, and we have some reason to think that immediate action is required to avoid catastrophe, the precautionary principle states that we should take that action. Some argue for the precautionary principle on the grounds that we have an ethical obligation to avoid catastrophe, whatever the practical costs. There is much to say in favour of ethical arguments. Still, some people will remain unmoved by them. These arguments will appeal only to those who accept the principle that ethical considerations may override considerations of costs and benefits. Nonetheless, there are, as well, purely prudential reasons in favour of the precautionary principle that should, I will argue, convince those not already persuaded.

Some authors have argued that if normal canons of rationality cannot recommend maximum caution, then we should adopt a rival rationality that places more emphasis on democratic and ethical values. Others have pointed out that decision-making about global risks has been mostly a matter of ideology anyway, so we may as well openly discuss these values and make decisions for political, cultural, and ethical reasons rather than on the pretence of objective scientific grounds. I will critique this view and reject democratic epistemology for its failure to distinguish between epistemic values and action-guiding values. I think that we can admit the value-ladenness of science without moving to the position that we need to change our conception of rationality.

Nonetheless, I will argue that precaution is the rational course of action, even independently of ethical considerations. Using arguments parallel to those of Blaise Pascal and William James, I will argue that the prudential reasons for precaution are overwhelming and should convince those not already persuaded by ethical arguments. While the models discussed in part I can only reveal possible futures and not probable futures, the catastrophic threats posed by such things as global warming, ozone depletion, or population increase, though not firmly established, represent what James would call "live options"; that is, they present us with a possibility that is at least plausible, and they force us to make momentous decisions. I conclude that we cannot afford to risk catastrophe despite the high costs this decision will involve.

PART ONE

Predictions of Global Catastrophe Based on Current Models of Global Systems Are Uncertain

· 1 ·

Four Models of Global Systems

Scientists have turned their attentions to computer modelling as a method for studying global systems in order to evaluate claims of potential catastrophe and to see whether intervention is possible. In chapter 2, I will discuss the nature of modelling in general. In this chapter, I will sketch some well-known models of global systems designed in response to fears of global catastrophe and supply some of their history.

The models I am interested in each deal with a unique problem. These models of global systems examine the potential of some human activities to result in catastrophe. Nuclear winter, ozone depletion, overpopulation, and global warming are potential catastrophes brought about by human behaviour that are global in scope and have potentially irreversible consequences. Unlike models of local systems, these global models cannot be fine-tuned by trial and error. We cannot perform experiments on global systems to see whether the predictions made by models of nuclear winter are accurate. The unique nature of the dilemma is that we need to decide whether to act to prevent these potential catastrophes before we have strong evidence whether our fears of them are even realistic. Scientists hope that the results of the modelling exercise can be used by policy-makers to guide human action.

NUCLEAR WINTER

If the direct catastrophic consequences of a nuclear war were not bad enough, various scientists were warning in the early 1980s of the possibility of a new and previously unimagined disaster: nuclear winter. While the radioactive fallout and breakdown of organized society resulting from a nuclear war would certainly be horrific, the end of all life on the planet was not usually included among the consequences.[1] However, the ensuing global climatic effects might indeed ensure the apocalypse. Computer modellers conjectured that in the aftermath of a nuclear exchange, such a large quantity of soot and dust particles would be dispersed throughout the atmosphere that they might change the climate. The dust cloud might reflect so much sunlight back into space for such a long period of time that Earth would be plunged into a winter that would have catastrophic effects on the planet's life. Thus, it was urged that the global nuclear stockpile be reduced to a level below that which would trigger this potential catastrophe.

One of the first, and certainly the most influential, of the nuclear winter models was published in 1983.[2] Its simplicity was part of its persuasive force. The calculations were based on a representation of the atmosphere as a 1 cm^2 column of air, and the only variables considered were the area, the quantity of fuel in that area, and the amount of smoke that would be produced by burning. The authors claimed that the model was partially inspired by an analogy with volcanic eruptions and Martian dust storms. Study of these phenomena revealed that there are two simple mechanisms by which dust, soot, and sulphuric acid aerosols cool global temperature. First, these particles directly lower the global temperature by reflecting sunlight and thus reducing the amount that reaches Earth's surface. Second, they indirectly cause cooling by acting as condensation nuclei in cloud formation, thus further blocking sunlight. This simple model ignores regional geography; the effects of the wind in transporting the smoke around the globe; the heat capacity of the oceans; the effect of possible coagulation of particles; the altitude of the smoke cover, as well as whether it was a continuous blanket or in patches; and other complicating details. It is nevertheless able to explore the basic mechanisms behind the temperature changes expected from different amounts of particles blocking sunlight.[3]

Other models followed from researchers around the world. Unfortunately, the range of these predictions was far from precise. Tony Rothman reports that one model showed a potential attenuation factor of sunlight in the range of 2 to 150 times, while other models predicted temperature drops of 35, 20, and 8°C.[4] Nonetheless, each of these modellers regarded the others as confirming their own findings.

Besides lack of precision, there were many other problems with the models. Many of the predicted effects were based on an analogy with the behaviour of the sulphuric acid aerosols released in volcanic eruptions – but these would not be present in the nuclear winter case and there was no evidence to support the assumption that soot and dust would behave in the same way as the sulphuric aerosols.[5] In the USSR Academy of Sciences model of Aleksandrov and Stenchikov, no distinction was made between dust and soot with respect to their different reflective properties; this resulted in an overestimation of the nuclear winter effect.[6] There were large uncertainties with regard to the quantity and "blackness" of smoke that would be produced, as well as how much would be immediately washed out by rain (estimates ranged from 10 to 80 percent).[7] There was disagreement over whether there was a minimum threshold level implying that a relatively small burn area would trigger the full nuclear winter effect. (This concept was central in arms policy considerations with regard to desirable levels of total world stockpiles and the advisability of limited strikes.)

Despite their problems, nuclear winter models gained a great deal of support. The rapid acceptance of the nuclear winter vision might have been a result of its powerful imagery, which gripped anyone who heard it in a poetic, but nightmarish daydream. However, in his essay "A Memoir of Nuclear Winter," Rothman argues from personal experiences and discussions with some major participants that this widespread acceptance of a suspect model was a result of the noncritical stance that most scientists adopted for political reasons. (I will discuss these sorts of reasons for accepting predictions of global catastrophe in part 2.)

Later, more complicated models were designed, and the conclusions had to be modified to allow for the mitigating effects mentioned above. It turned out that the nuclear winter models were not "robust"; that is, small changes in the initial conditions and assumptions would result in wide variations in the predictions of temperature decreases and duration of the dust cloud. Since estimates of the amount of smoke produced by nuclear explosions are very uncertain,[8] this sensitivity of the model to small changes in assumptions is a serious defect. For example, if the nuclear exchange takes place in the winter season instead of the summer season, the temperature drop is smaller by one order of magnitude.[9]

S.L. Thompson and S.H. Schneider argue that, as a result of the above considerations, the deep-freeze interpretation of nuclear winter is now dismissed by most scientists as unlikely. Instead, these authors suggest that we think in terms of a nuclear autumn rather than a nuclear winter.[10] While not suggesting that nuclear war will have no other lasting environmental effects, Thompson and Schneider nonetheless argue that "on scientific grounds the global apocalyptic conclusions of the initial nuclear winter hypothesis can now be relegated to a vanishingly low level of probability."[11] That is, there does not seem to be the potential for human extinction resulting solely from the climatic change that would follow nuclear explosions. The *possibility* of nuclear winter still exists, but the revised models suggest a much lower *probability* than the original models, as well as the absence of any minimum threshold.

OZONE LAYER DEPLETION

What is interesting about the history of models of ozone layer depletion is that the fear they generated was sufficient to initiate worldwide preventive action *before* any consensus was reached about the plausibility of the models and before there was any measurable damage.[12] In this case, the causal evidence was not discovered until afterward. Action was taken to reduce the use and release of CFCs (chlorofluorocarbons) as early as 1978. However, it was not until 1988 that a study of the Antarctic atmosphere was published that revealed that there was a direct correlation between increased concentrations of chlorine and decreases in ozone.[13]

The first publications that implied that CFCs might be damaging the ozone layer came in 1974.[14] CFCs migrate up to the stratosphere where they are broken down by short-wave ultraviolet radiation. As the CFCs break down, the chlorine released acts as a catalyst in an ozone destroying reaction that takes place on ice crystals created in large polar vortices. A single chlorine molecule can destroy about 100,000 molecules of ozone.[15] As the ozone thins, the amount of UV-B radiation reaching Earth's surface increases, and this results in an increased frequency of skin cancer and retinal burn, the suppression of immune systems, decreased agricultural yields, and even, perhaps, the death of the oceans, as UV-sensitive plankton and other creatures low on the food chain are destroyed.

The effective political actions of a strong environmental movement led to preventive action being taken as early as 1978, with a ban on aerosol sprays in the United States, Canada, Norway, and Sweden.[16] However, the production of CFCs for other uses (refrigeration, thermal insulation) continued to grow. It was not until 1984 that the ozone hole over Antarctica was first noticed. A 40 percent decrease in ozone over ten years had gone previously unnoticed because researchers had rejected low readings as anomalous.[17]

The nature of the phenomenon was yet to be understood. Explanations and evidence would not come until three years later. Nonetheless, international political agreements were reached in 1987 to limit production and use of CFCs. An international agreement to reduce CFC use, known as the Montreal Protocol, was signed in 1987. Soon after, the results of a crucial experiment were published. A NASA research plane had flown over the Antarctic taking measurements of the relevant variables. The measurements revealed that decreases in ozone concentration were directly proportional to concentrations of chlorine present.[18]

OVERPOPULATION

The population of Earth has just reached approximately six billion, and it is still growing. While it is clear that we must reflect on the potential consequences of our current rate of population growth, it is not obvious to everyone that the growth itself must soon end. On the one hand, there are those who believe that our resources are very limited and, therefore, that we must take preventive measures to limit growth now, or else expect the much less desirable checks of massive famine and global economic collapse. Others maintain that we have enormous supplies of cheap energy – the sun or the wind or uranium – and vast quantities of untapped metal resources on the ocean floor and in Earth's core. Even if these resources are ultimately limited, there is still room for a huge increase in population. We are often reminded in these arguments that, because the added population also contributes to production, an increase in population is not simply a drain on our finite stock of resources – the more the merrier, so to speak.[19] The point of this argument is that the globe's carrying capacity cannot be described as an absolute limit that is independent of other contexts. In the words of L. Sjoberg, "Carrying capacity is a socially organized threshold and not a simple technological phenomenon."[20]

One of the most influential arguments in support of the conclusion that we have exceeded, or will very soon exceed, Earth's carrying capacity is to be found in Meadows et al.'s *The Limits to Growth* (and later in Meadows et al.'s *Beyond The Limits*). The claims made there are based on a computer model of Earth that attempts to take into account the interactions of several systems. This "system dynamics" model characterizes the planet in terms of several components – population, pollution, resources, food production, and other interrelated subsystems. First it gives descriptions of the behaviour of each subsystem (for example, population grows exponentially in certain circumstances) and the connections between these component systems (for example, supply and demand laws, the effects of pollution on agricultural yields, delayed reactions, and any known positive and negative feedback loops). These descriptions are a mixture of theoretical assumptions and empirical measurements. The model then makes countless calculations in an attempt to discover the dynamics of the whole system under different initial conditions obtained by varying assumptions about such matters as resource availability or technological efficiency.

The authors identify four possible futures that would result under various assumptions.[21] The population and economy could (1) grow continuously, (2) slowly approach a limit and then stabilize, (3) overshoot the limit slightly and then oscillate above and below that limit in a series of minor crises and recoveries, or (4) overshoot the limit and "collapse" inelastically. The authors reached the conclusion that, if the world's systems continue to operate in the same manner as today, then the fourth future they describe will be our future – the world will overshoot its limits within one hundred years and collapse irreversibly.[22] Desertification, depletion of fisheries, loss of species, and losses in industrial capacity will be so sudden and severe that recovery, if it occurs at all, will take decades at least. (In fact, they claim we have already overshot limits in several areas.[23]) This catastrophic future will result because the indicators of problems are delayed (for example, it takes about fifteen years for CFCs to migrate to the stratosphere, where they start to break down ozone[24]), responses to these problems are often slow, and many subsystems are unable to recover quickly, if at all, after large setbacks.

The model is not specific about how global collapse will come about. It can come in various ways – depletion of natural resources or economic failures, for example. Unfortunately, if one of these paths to destruction is avoided, another one is precipitated. Diverting financial resources from polluting industries will, given one scenario, lead to economic collapse. So too will boosting agricultural production in order to feed a growing population. Thus, the model reveals that the problem is not simply one of physical limits to the world's finite natural

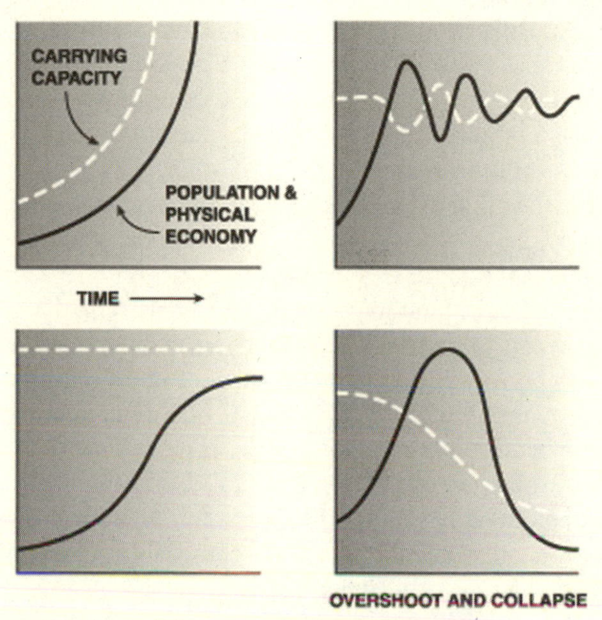

Four modes of population approach to carrying capacity
Based on figure 4-2 in Meadows, Meadows, and Randers, *Beyond the Limits.*

resources but rather one of limits in what they call the world system's "ability to cope."[25] The writers are claiming that their model reveals that our future is threatened by population growth not simply because resources are scarce but also because of the inherent and inescapable problems arising from the structure and organization of our world system. Thus, the writers claim, even with very optimistic assumptions about increased economic and technical efficiency, recycling, improved agricultural yields, and pest control, the world system will still suffer a collapse under the weight of growing population.

GLOBAL WARMING

Recent climate models built to study the threat of global climate change imply that should present patterns of human behaviour persist and the concentration of carbon dioxide, methane, nitrous oxide (greenhouse gases) continue to increase over the next hundred years, the average global near-surface temperature will increase 1.4-5.8°C. The Intergovernmental Panel on Climate Change (IPCC) estimates that the global mean temperature will increase approximately 0.1- 0.2°C per decade and before the year 2100 will be at least 3°C above current averages.[26] The stratosphere, meanwhile, will undergo a cooling effect. Sea levels might rise as a result of both thermal expansion of the water and increasing run off from melting ice caps. The interior of some continents will become more arid and the frequency and severity of hurricanes and typhoons could increase.[27]

Some observers argue that the average global mean temperature has already increased by 0.3–0.6°C over the last hundred years. If they are right, a plausible explanation for this increase is the action of the greenhouse gas warming effect represented in these models. On the other hand, the existence of a statistically significant increase is disputed,[28] and the increase is not outside the limits of natural variation anyway.[29] The climate cycles naturally on several different time scales as well as fluctuating randomly; for example, sunspot cycles and the El Niño effect are both major factors in determining global climate. There are also long-term balances and energy shifts between the various global components. The natural variability of Earth's climate is determined by changes in ocean circulation, atmospheric dust, and long-term fluctuations in Earth's "wobble" (Milankovich cycles) as well as changes in solar radiance. Because the temperature increases predicted by global climate models are, so far, too small to distinguish them as outside the natural limits, any evidence for such increases is incapable of confirming global warming theory. The influence on climate of natural system cycles continues to be much greater than that of human behaviour.

Even an absence of a measurable global warming will not falsify the greenhouse hypotheses, however. The global climate is a complex system of energy transfers involving cloud formation, air currents, ocean currents, changes in rainfall and soil moisture, and storm frequency – as well as temperature change. An absence of global warming to date, then, does not necessarily mean that the climate is not being adversely affected by greenhouse gases. The climate change might be occurring in one of these other areas of the system.

To study these energy transfers, much effort has been spent in developing computer simulation models of Earth's climate. These "general circulation models" represent Earth's climate with a series of mathematical equations of the processes occurring in various spheres of Earth – the atmosphere, the oceans, the continental landmasses (the geosphere), the biosphere, and the cryosphere (ice caps and glaciers). The behaviour of each of these areas varies considerably with respect to speed and complexity; for example, the geosphere and the crysophere affect climate greatly, but change very slowly, while the atmosphere and ocean surfaces can change rapidly. Models must capture this complexity for accurate representations.

Some components are incredibly complex. Consequently, modellers make a large number of simplifying assumptions about them. Particularly difficult to model are cloud formation, biosphere/climate interactions, and ocean currents; consequently, researchers remain largely uncertain of their behaviour. Since modelling the details of small-scale processes like those involved in cloud formation and ocean currents is also limited by computer power and expense, it is not attempted. Instead, these processes are treated as "black boxes" and statistical observations are inserted. This procedure is known as parameterization. Thus, for example, modellers collect data on general correlations between humidity and temperature levels, and assume that they are sufficient for modelling purposes. Thus, the need for detailed equations representing the physics of these relations is avoided. For example, while the Canadian Climate Centre model has the highest spatial resolution of any model, it does not include ocean currents.[30]

A major methodological objective of global circulation models is to represent the present climate – including seasonal changes, evaporation, and rainfall – and then to adjust the components of the model to represent changes in greenhouse gases. (A common model scenario assumes a doubling of carbon dioxide (or equivalent), while other models adjust the increase more gradually.) Unlike that of regional weather forecasting, the goal of climate modelling is not to predict the system's behaviour at any particular point in time but only the general behaviour of the system. Thus, the resolution of a climate model is intentionally lacking in detail. Global circulation models represent the atmosphere as a series of grid points on three-dimensional axes. The number of points is largely determined by computer power, because for each point several meteorological variables are calculated. For detailed, regional weather forecasting, these grid points are in close proximity. In contrast, for global modelling, the required resolution is much lower, and grid points may be several hundred kilometers apart with only 10–20 vertical levels.[31]

One climatologist has claimed, ironically, that "the greenhouse effect is the least controversial theory in atmospheric science."[32] What he means is that, while there are many uncertainties about whether global warming will be catastrophic, the general behaviour of the components in the model is not disputed. We know, from the analysis of ancient air found in bubbles trapped in glacial ice cores, that the concentration of greenhouse gases has increased. The warming effect of these gases is well understood. The ultraviolet rays of sunlight penetrate the atmosphere, strike Earth and are reflected back as infrared rays, which are then absorbed and trapped by the greenhouse gases in the atmosphere. We know that the greenhouse effect is what makes this planet habitable, and that the controversy is really about whether this effect will be increased and, if so, whether this increase will be detrimental to humans. We also know that some historical climate changes – for example, the end of the last ice age – are correlated with changes in greenhouse gas concentrations. (However, this correlation does not imply causation.)

The controversy lies in whether average global temperatures have already actually increased or will increase. The rate of change is also highly uncertain and yet crucial for ascertaining whether global climate change is a serious threat. The theory of the greenhouse effect is uncontroversial in the sense that everyone agrees that an increase in the concentration of greenhouse gases will affect the climate somehow. What *is* controversial about the theory is just what the consequences will be: how much warming and where? Will there be any negative feedback from the atmosphere, the oceans or the ice caps to mitigate any warming?

Evidence cited for the increase in global temperature consists in the recorded warming trend in recent global meteorological history (the hottest years on record have been in the last two decades), the observation that the average surface temperatures have increased by 0.3–0.6 °C in the last hundred years, the fact that Canadian lakes have an increasingly long ice-free season, and many other changes in the timing of plant and animal cycles such as flowering and egg-laying.[33] However, there is no evidence that these changes are directly linked to anthropogenic emissions; it might still be possible that the temperature increase is just part of normal global temperature variations.

Some climatologists who believe that Earth is warming claim that it is unlikely that such a warming trend could have occurred merely by natural processes: "with 99 percent confidence we can state that recent global warming is a 'real warming trend,' not one occurring by chance or accident."[34] Others, however, are not convinced: "recent warming is still within – although perhaps pushing – the upper limits of natural variability."[35] Even if there is a warming caused by anthropogenic emissions, its magnitude is not yet outside natural limits, and thus the existence of this warming cannot serve as evidence for the increased greenhouse effect hypothesis.

Even if the prediction of increased warming is accepted, however, it is not *obvious* that the predicted three or four degree temperature increase presents a serious problem for humans and other creatures. Warming might even benefit some regions without harming others. Winters and nights could be warmer without necessarily increasing summer or daytime temperatures. The growing season might be extended at high latitudes, and high carbon dioxide concentrations increase growth rates of many plants while simultaneously decreasing their need for water, thus making agriculture possible in semi-arid wastelands. Previously lifeless areas might become capable of sustaining ecospheres. Such changes might result in benefits in agriculture and forestry.[36]

Others argue that this optimism is too hasty. First of all, complex systems cannot be changed that simply. For example, even though the growth rate of plants might increase, pests too are likely to be more prolific. Though a slight increase in temperature might be beneficial in some areas were it to come on gradually, if it were to come on rapidly, ecosystems would not be able to adapt. Furthermore, even small changes in global temperatures can change global climatic *patterns*, resulting in increased frequency and severity of hurricanes, droughts,[37] and other weather disasters – what Bernard calls "climatic stress."[38] Thus, even a small temperature increase might present a serious problem. Analogies with past abrupt climatic changes – cooling at the end of the dinosaur era, temperature fluctuations due to enlargement and diminishment of ice-age glaciers, the 1930s drought in the US Midwest – indicate that a shift of only a few degrees can have severe consequences.

Second, the predicted increase of 1.4–5.8°C over the next hundred years is a global *average*. Climate models aggregate the climate of various regions into one large grid and thus ignore regional differences. However, it is the larger temperature variations within the various regions that are the source of the environmental anxiety. The poles will get warmer, perhaps resulting in some melting of the ice caps and flooding of the world's coastal settlements, endangering an enormous number of lives.[39] Even as coasts are flooded, the interior of some continents will be stricken with drought, straining irrigation resources, harming wetlands, and increasing wind-erosion of the land.

Uncertainties are not confined to regional variation, however. The chain of consequences triggered by an increase in average global temperature might result in either an acceleration or mitigation of the warming trend. It is possible that the effects of some process act in such a way as to enhance that process. This is known as a positive feedback loop. For example, a positive feedback loop might be that warmer oceans would absorb less carbon dioxide, or they could release tons of the greenhouse gas methane stored in sea floor mud or permafrost, thereby causing further warming.[40] Either way, the warming effect could be enhanced by about 0.5 °C per year.[41] Negative feedback loops, on the other hand, serve to reverse or slow down a process. In this case, it might be that an increase in global temperature would result in more bright cloud formation, which in turn could mitigate the warming effects by decreasing the amount of sunlight that reaches Earth. Once again, however, the pattern of cloud formation would have regional variations and would have different effects in different areas.[42] Clouds of ice crystals over the Arctic behave differently than clouds over the tropics with respect to how much heat is trapped or reflected.[43] The effects of clouds vary greatly with their brightness and structure. Clouds can deflect light and thus decrease warming, or act as an insulating layer, thereby increasing global warming. Cirrus clouds, which now cover approximately 16 percent of Earth, act as a blanket and can have a net warming effect, while marine stratocumulus clouds, which now cover approximately 34 percent of Earth, have a net cooling effect because their brightness reflects sunlight back into outer space. Slight differences between models in the way that the dynamics and distribution of these clouds are represented will result in large differences in predicted warming.[44] We also do not know whether the concentration of greenhouse gases will continue to increase or whether the deep ocean currents will absorb most of the greenhouse gases (although it is suspected that the rate of deep water absorption will decrease[45]). We think there must be a carbon sink somewhere because of the missing carbon problem. The concentrations of carbon dioxide in the atmosphere are not as high as estimates of carbon dioxide emissions, and we do not know where the missing carbon is going.[46] Other negative feedback mechanisms that would limit the effects of global warming have also been postulated. Polar ice caps might actually *increase* in area. Although the poles might get warmer, it might still be cold enough for snow. Higher average temperatures mean higher amounts of evaporation of water, and this could lead to a greater accumulation of snow. This increased area would mean a greater reflectivity of the sun's energy. This, in turn, would slow the warming.

Another complication that modellers must consider is the effect of sulphate emissions. The burning of fossil fuels, which is the major source of greenhouse gas emissions, is also a source of sulphate particle emissions. Sulphate particles in the atmosphere initiate a cooling process by reflecting light.[47] Researchers estimate that approximately 25 percent of the global warming effect can be mitigated by the presence of sulphate particles.[48] Ozone reduction might also indirectly cause cooling through complicated atmospheric chemistry leading to increased cloud formation.[49]

Modelling oceans is also complicated. Increasing water temperatures will change levels of salinity, biological activity, rates of carbon exchange between surface and deep currents, and much else.[50]

How these feedback loops will work is the largest source of uncertainty in global circulation models. Stephen Schneider concedes that "present models, crudely reproducing only average cloudiness, can say little that is reliable about cloud feedback – or about the many other feedbacks."[51] Similarly, an IPCC assessment of general circulation models concludes that "in their current state of development, the descriptions of many of the processes involved are comparatively crude."[52] (Although the most recent IPCC report claims that models are improving all the time.[53]) Even well-known feedback mechanisms, such as those between climate and vegetation, are not always included owing to limited computer resources.[54]

In the case of greenhouse models, there will be no one discovery that will confirm or falsify the models. In the ozone layer depletion case, as we saw, a single experiment confirmed, to the satisfaction of most researchers, the connection between chlorine concentrations and ozone depletion. However, in the global warming case a large number of climate variables is involved; the unfolding of global warming will not be a simple rise in average surface temperature. Confirmation of models that predict global warming will be a slow process of gathering data about many parts of Earth's climate over many years. (For example, some models predict an increase in precipitation above 30°N latitude, and a decrease below this latitude.) Research is ongoing, but it appears that some new studies hint that global warming will not be as great as was previously thought.[55] For example, currently estimated increases in sea level are not nearly as large as those once thought likely.[56]

Meanwhile, the actions recommended to avert this possible catastrophe will be much more costly than were required in the ozone-depletion case. Greenhouse gases are produced by routine and fundamental practices in transportation, industry, and agriculture. Reduction of emissions in these areas will require a much larger trade-off than was required in reducing CFC use. According to one estimate, even if we were to replace every coal and oil-burning electricity generating plant with one that produces no carbon dioxide, total carbon dioxide output would be reduced by only 25 percent.[57] Furthermore, because carbon dioxide resides in the atmosphere for long periods, even with a complete cessation of emissions today, about half of the recent increase in concentration would still remain in the atmosphere in one hundred years.[58] To stabilize greenhouse gas concentrations at current levels, reductions in use will have to be dramatic. In his book *The End of Nature*, Bill McKibben comments that "the sacrifices demanded may be on a scale we can't imagine and won't like."[59]

·2·

What Is a Scientific Model, and What Do We Do with One?

In the previous chapter, we acquainted ourselves with four contentious models of global processes. In the next chapter, we will scrutinize the same four models to see whether any of them generate predictions reliable enough to support decisions about economic or political policies. In this chapter, we will try to understand the nature of models and how models are used in science. What is a scientific model, and what do we do with one?

The main function of scientific models is to provide predictions and what Mary Hesse calls "intellectually satisfying" explanations.[1] A scientific model is a representation of the workings of some system, a representation meant to explain the observed course of events and to provide predictions. The ultimate goal of a model is to supply enough understanding for us to be able to intervene in the system to achieve some end. We might model a traffic system in order to guide decisions about the best use of stop signs or one-way streets with the goal of obtaining safer and more efficient traffic flow. We might model the population dynamics of a wilderness system with the goal of preserving the elements of that system necessary for the survival of some species. We create a model in the hope that it may provide a good estimate of the time scales involved in a system, reveal general behaviour patterns in it, and identify the most sensitive points from which to influence that behaviour.

Hesse distinguishes three types of scientific models – physical models, analogies, and mathematical models. Some models are physical representations – for example, a wind tunnel containing a scale replica of an airplane or a mechanical model of a water-walking lizard used to discover the forces involved in the animal's locomotion.

Analogies draw attention to similarities between the thing being modelled and something we are already familiar with. Our explanation of light is modelled on our understanding of waves, and kinetic energy is explained through an analogy to billiard-balls. The potential effects of global warming have been described by analogy with the drought in the US Midwest in the 1930s,[2] and the first nuclear winter model was based on an analogy with the cooling effects of Martian dust storms.[3] These supposed similarities will, modellers hope, explain the behaviour of an unknown system in terms of the physical structure and laws of another, better-known system and serve as a guide for making predictions about that behaviour.

Hesse's third type of model is a mathematical or probability model. Models of this type consist of sets of equations purporting to correspond to some features of the world. It is this last type we are especially concerned with when discussing models of global systems. Mathematical models typically consist of a network of equations describing the behaviour of the system under study, and they enable the modeller to make predictions about the future behaviour of that system with respect to certain variables. Mathematical models are not intended to provide precise predictions about what will happen at a certain time in the future. Rather, they aim to uncover general patterns and trends that might arise as a result of gradual changes in the system – say, increases in carbon dioxide concentration or population. Supercomputers are now widely used to examine the complex direct and indirect relationships among components of the global environment. These models aim to simulate the behaviour of a large system over long periods of time and supply predictions about what might happen when various components of the system are changed.

We know from experience that some models are successful at both explanation and prediction. We might, for example, construct a model of traffic circulation that can be adjusted in various ways. In the model, we could adjust the timing and synchronization of stoplights, or increase speed limits, or restrict the direction of traffic flow on main arteries. By examining how the model behaves with these changes we are able to make predictions about what will actually happen in some city where these changes are made. Traffic models can be tuned by trial and error until they are successful. We now have enough experience with this kind of model to believe that the predictions will be fairly reliable, and that changes in traffic patterns will result for the reasons that the model says they will.

When a model is unable to give us one of either explanation or prediction, our warrant for accepting the model will be tied to the remaining property. If the model is successful at prediction but does not contain an internal structure that explains events, then our acceptance will be based on predictive success alone. If, for some reason, the model is unable to supply testable predictions, then our warrant for believing that this is a good model will be tied solely to the cogency of its structure and explanations.

Some computer programs forgo the attempt to represent an internal structure that explains events and are designed solely to make accurate predictions.[4] While representational models provide details of the structure of a system and are intended to explain as well as predict, nonrepresentational methods of prediction – so-called expert systems – simply collect correlations between measurements of features of a natural system. For example, in global climate models, many complex interactions among clouds, oceans, and precipitation might be simply translated into a mathematical relation between temperature and quantity of precipitation. A scientist notices a correlation, say, between the temperature of the air over some region of the ocean during rainstorms and the amount of precipitation in those storms: the lower the temperature, the smaller the amount of precipitation. The scientist then attempts to find a function that describes this correlation, so that given the temperature during some storm, the amount of rain can be computed. Of course, this expert system gives the scientist no insight into why the function discovered works to generate successful predictions, for nothing in it is described either as cause or as effect, nor does the system supply any guidance as to how to intervene in the world to alter the amount of rainfall. Constructing functions that describe correlations in measurements between features of the world is called parameterization. The quantitative relations between variables is captured, but in such a way that no explanation of why the two variables are related in the way they are related is possible.

Nonrepresentational computer programs are simpler and usually make more successful predictions – because they are dedicated to this purpose. Obviously, this approach will not be able to supply detailed causal explanations of the system's behaviour, and this might be just what is needed for us to understand the system sufficiently to intervene in desired ways. Representational models have a structural component that supplies an explanation of physical behaviour rather than simply describing relations and correlations. Too much detail is undesirable, however. Modellers must make several simplifying assumptions in order that models can be formulated without getting bogged down in details that may be impossible to calculate. Even if these details were available, the sheer volume of data required for detailed predictions could make the models unwieldy. The job of a modeller is to start from general equations of fundamental principles, say, of economics or thermodynamics of climate, which are believed to hold true in any situation, and then to add enough detail to examine a particular issue. A skilled modeller must know what to leave out as well as what to include. Karl Popper's maxim about simplicity in science is especially applicable to modelling: "Science may be described as the art of systematic over-simplification: the art of discerning what we may with advantage omit."[5] Good scientific models must strike a balance between generality, realism, and precision.[6]

Some components of the real system, then, must be assumed to be irrelevant or independent of the particular physical behaviour we are focusing on. When modelling Earth's tectonic plate activity, for instance, one need not include "hot spots" like Hawaii because these anomalies are not central to the model. Other components cannot be ignored but, nonetheless, must be vastly simplified. As mentioned above, the global circulation models used to examine global warming, for example, do not attempt to represent the detailed behaviour of cloud formation or ocean currents – even though these are extremely important. Instead, these processes are treated as black boxes with the details eliminated. Modellers assume that it is not necessary to mention clouds in order to describe the relationships between temperature and rainfall or, to give another example, that is sufficient to describe the rate at which ocean currents distribute heat around the globe without giving descriptions of any particular currents. Thus, purely predictive components are incorporated into a general structural, explanatory model.

Surprisingly, perhaps, the two functions – prediction and explanation – are often in opposition to each other. R.H. Peters argues that the more explanatory a model, the less it tends to be successful at furnishing accurate predictions.[7] This is because explanations reflect the various commitments the modellers have toward, say, reductionism or various background theories, and these commitments are not easily sacrificed for the sake of stronger predictive power. Thus, a scientist might prefer a model with intuitive plausibility to one without this plausibility and slightly more predictive success.

The trade-off between prediction and explanation is reflected in the debate between advocates of complicated, three-dimensional global climate circulation models, and advocates of simpler models such as energy balance models.[8] On one side, it is assumed that the accuracy of the representation is proportional to the quantity of realistic detail, and only a realistic model will provide the explanations that are required to intervene in the real system. They might argue, for example, that feedback processes can be predicted only by those global circulation models that include many complicated details about winds, eddies, currents, evaporation rates – the component processes of essential dynamics and thermodynamics of the feedback. However, including these details for the sake of greater realism can make for a very complex model, one difficult to use.

Simpler models, however, might focus only on generalities such as the difference between energy entering the climate system from solar radiation and energy exiting the system. Advocates of these simple models claim that they are better at, or just as good at, predicting climate change.[9] They argue that the reduction of large scale behaviour to component processes has not been successful; for example, the simulation of feedback is extremely sensitive to small changes in the equations representing particular component processes on this small scale. Scientists can never be sure that they have the details right, and every adjustment results in different dynamics of feedback. Thus, they argue, modellers might actually be weakening the model by adding more complexity, making it less able to simulate more general interactions. The compromise, however, might be acceptable to those who desire an explanation of climate change and not just a prediction.

Nancy Cartwright, too, has argued that explanatory power comes at the cost of descriptive adequacy.[10] Her "simulacrum" account of explanation makes models an intermediary between theoretical laws and reality.[11] According to Cartwright, models, as the name of her account implies, are a mere semblance of the real thing. They are, she claims, false simplifications that are constructed according to theory and then used as general frameworks to generate more specific explanations for the phenomena under investigation. Models are fictitious descriptions that modellers tune with ad hoc adjustments until they agree with the phenomena. There are two ways to do this. Either the model itself can be adjusted to fit the phenomena or the descriptions of phenomena can be "prepared" to fit the model.

Models are idealized, for example, by assuming frictionless surfaces, or point sources of gravity. Idealization makes the model applicable to many different cases. Take, for example, a simple model of a mass-spring system, a model of the sort first-year science students encounter in their physics texts and labs. In this model, the force on the mass is proportional to the amount the spring is stretched. Cartwright points out that an "oscillator" model of this sort can be used to represent pendulums, hydrogen atoms, and lasers–provided that certain idealizing assumptions are made.[12] The generality of the model provided by idealization gives the model its explanatory power. The model supplies a common explanation for multiple cases.

Alternatively, Cartwright argues, descriptions are often prepared to fit the model. In an essay entitled "Fitting Facts to Equations," she explains how accurate descriptions of phenomenon are not always amenable to mathematical analysis.[13] These complex descriptions must then be "fictionalized" in ways that make them match available equations.

On the standard, covering-law account of explanation, one assumes that the fundamental laws of a theory are true of reality; that is, phenomena can be explained by reference to general theoretical laws. Cartwright argues that this is a bad account of explanation that gives the false impression that theoretical laws "govern" reality. The simulacrum account of explanation that she defends, on the other hand, claims that the fundamental laws of a theory are true of objects in models only. The models, in turn, serve to fit reality into the theory by way of approximations.

The important implication of her account is that there is no single correct explanation for any particular phenomenon, but, rather, an indefinite number of models that make different approximations for different purposes. Some models will emphasize accuracy and thus be designed to provide precise calculations and predictions. However, other models will sacrifice accuracy in order to explain a wider range of phenomenon with the same model. Some models will be designed to calculate global averages, while others will attempt to reveal regional details. Different models will focus on different things for various purposes, and since these models will often be incompatible, we will not be able to choose between them simply by checking which ones better fit the facts.[14]

Ronald Giere also endorses an account of explanation that emphasizes the role of models. Giere arrived at his account of scientific explanation by empirical investigation of the way scientists actually use models. He studied introductory physics textbooks and their models of mass-spring systems, pendulums, etc. and discovered that most models are simplified with the specific goal of studying only one variable at a time and are deliberately tailored to ignore all the complicating factors. It is the set of all these "judicious approximations" that constitutes a mathematical model.[15] Thus, it is more correct to talk of a collection of models for a system, each with a separate purpose, rather than of a single model.

These two accounts of explanation parallel what has been called the "semantic view" of theories.[16] In the semantic view, theories are formal systems, and a model is any instantiation that satisfies the equations of that system. In formal terms, one could say that a theory is an "uninterpreted" system with "open" variables and a model is any "solution" or "interpretation" of this system – i.e., the variables are assigned values. Since the theory may have several instantiations, there will be several models. This is a *semantic* view because an open theory does not acquire a *meaning* until it is interpreted. As a string of uninterpreted symbols, a formal system is merely a syntactic structure that does not refer to anything. The important implication of this view is that models are not to be understood as sets of statements, and, thus, a theory is not restricted to a single interpretation, but rather encompasses a family of models. Giere writes: "Since models are nonlinguistic entities, they can be characterized in many different ways, using many different languages. This makes it possible to identify a theory not with any particular linguistic formulation, but with the set of models that would be picked out by all the different possible linguistic formulations."[17] If models are not linguistic formulas, it makes no sense to describe their relationship to the real world as either true or false. Instead, Giere argues that the relationship is one of "similarity," which must be specified in terms of *respects* and *degree*. The various models of a cluster are similar to the real world in various respects and to various degrees. A particular model is similar in some respects to the real world, while a different model might be similar in other respects. A "cluster" or "population" of models can, thus, represent the real world.[18] A theory, then, is a collection of models together with hypotheses that relate these models to the world in various ways.

I will argue in the following chapters that models of global systems present us with a deep problem of evaluation, one that has serious implications for public policy. The global models we will now examine in detail predict that unless we change our ways drastically and soon, we will be crushed by catastrophe. As frightened as we are by the prospect of catastrophe, we do not want to change our ways drastically, or at least we would like to be able to phase in the needed changes gradually; so we would like very much to know whether we have good reason to trust these predictions. Generally the best thing to do with a new, rough model of some natural system is to test it against its predictions and then to correct it by the method of trial and error until it generates precise and reliable predictions. Only after such a process will responsible scientists declare a model sound and hand it over to makers of public policy. But we cannot wait for researchers to perfect doomsday models by revising them in light of the successful and unsuccessful predictions they generate, for to do that is, as things stand, to court catastrophe. We must, then, accept or reject these models and use them or scorn them, only according to their ability to give intellectually satisfying explanations of currently observable phenomena. If I am right in what I will now argue, though, we must seriously doubt that any of them do provide intellectually satisfying explanations of the phenomena they are supposed to explain. And yet, one or another of them might very well be an accurate model of how things are, a model that we scorn at our peril.

· 3 ·

Assessing Models of Global Systems: Direct Tests

It is clear from the examples given in chapter 1 that global models are porous with uncertainty. When, for instance, we consider fluctuations in solar radiance, complicated causal chains of feedback, and other variables, we find that global climate models are just as likely to underestimate global warming as overestimate it – and, therefore, might not be much good at all. How, then, are we to assess the predictions of catastrophe stemming from these models? Can we know (or have good reason for believing) that a model of a global system, from which predictions of the future are generated, is trustworthy?

Successful predictions are a direct test of a model's reliability, and I will first discuss criteria for assessing whether any particular prediction can serve as a good test of a model. The usual methods of tuning a model and assessing its reliability on the basis of predictive success are of limited use in the case of predictions of catastrophe, for the confirmation of predictions will come too late to be useful. The nature of the problem in question is that we are urged to take immediate action *before* predictive success is possible. Nonetheless, policy-makers would like to have some method of assessing whether models of global systems can make trustworthy predictions before recommending any action based on these models. We must, then, direct our epistemic energies toward seeing whether a model actually supports the predictions it is said to support. If the models predicting global catastrophe are right, we should probably be committing large resources or making large sacrifices in order to reduce any chance of catastrophe. Unable to use predictive success as a criterion for model acceptance, researchers must base their judgements on other criteria. Here, I discuss these other criteria. In the next chapter, I will discuss two general requirements for making predictions, arguing that these requirements are only partially met by models of global systems. In the rest of part 1, I will discuss indirect methods of assessing these models and conclude that current models of global systems leave us uncertain as to the likelihood of their predictions.

DIRECT TESTS

How is it to be determined whether a particular model has succeeded in its task of representing the behaviour it sets out to represent? What would constitute a good test that would allow the separation of good models from mere fanciful creations incapable of making reliable predictions? What reasons can there be that would make us have faith in some particular model rather than some other model?

Ronald Giere has summarized the elements of good tests for models, theories, and hypotheses in general.[1] A good test of a model, theory, or theoretical hypothesis, he argues, is supplied by a prediction that satisfies the following three conditions: (1) the prediction is implied by the model, theory, or hypothesis and the initial conditions, (2) the prediction is highly improbable unless the model, theory, or hypothesis is correct, and (3) the prediction is verifiable according to normal scientific standards.

In order for models of global systems to be testable they must generate predictions that are deducible from the model together with observations of the initial conditions. For global models these predictions might be in the form of statistical generalizations. The early nuclear winter models predicted a drastic reduction in global temperature in the aftermath of a limited nuclear exchange. Climate models are predicting an increase in the average global temperature in the range of 1.4–5.8°C in the next hundred years if production of greenhouse gases continues at the same rate. The "Limits to Growth" model predicts that near the year 2100, if current trends continue, there will be an inelastic collapse of agricultural and industrial production accompanied by a massive human death toll. Global models also make predictions of the *near* future. Global climate models predict that the northern hemisphere will warm more rapidly than the southern hemisphere, and the stratosphere will cool.

The second condition is that it be highly improbable for the predicted events to come to pass unless the model is correct. It is important, that is, that a successful result in the test not be a foregone conclusion. This is to ensure that not *any* model can pass the test. If a model predicts effects that many different models would all treat as likely to occur then the test fails to differentiate among these models. I will argue that climate models predicting increased global warming in the near future are like this; namely, the truth of the warming predictions gives little reason to prefer the climate model that explains that warming by reference to greenhouse gas, rather than, say, a climate model that explains the warming by supposing increased sunspot activity. The predictions do not differentiate between these two models because both predict the same thing. To conclude from an observed, slight global warming that that warming is caused by the increased greenhouse effect is to make a very weak inductive inference, given that other models and hypotheses also generate this prediction. Merely observing something consistent with a model does not justify the inference that the model that predicts that observation is trustworthy. What makes that move more compelling is the *improbability* of this observation unless the model were reliable. If incompatible models make the same prediction, then the degree of support that that prediction supplies to anyone is low. However, if the predicted state of affairs is unlikely on any other interpretation of events than that supplied by the model, then this makes it more likely that the inductive inference that the model is a good representation of actual events is a good inference.

Of course, models are not rejected or accepted on the basis of single predictions. Even if a model makes no single, improbable prediction, it might make several predictions that together satisfy the improbability condition. It is unlikely that a model could make good predictions in many areas without being a good representation, even when there is nothing remarkable about any one of its successful predictions taken just by itself.

To satisfy the improbability condition, it is important that the prediction be independent from the data used to construct the model in the first place. It is important, that is, that the test avoid begging the question. If a model was constructed to explain a certain event, then that event cannot be said to be a good prediction because it is not highly improbable – it was known, in advance, to be true. As Giere writes: “If the model was designed to use in explaining these results, what sense does it make to say that these results would be very improbable if the model did not apply?"[2] As an example of how this second condition works for simple models and hypotheses, Giere describes how Edmond Halley's prediction of the appearance of the now famous comet was based on Newton's physics and his own astronomical observations. Halley's theoretical model of the comet as a Newtonian system was tested by searching the historical records of previous comet appearances with similar orbits. Halley concluded that several recorded observations could be attributed to one comet that was in an elliptical orbit with a period of seventy-six years. This was a good test because this historical record was not used to construct the model.

So that the reader can appreciate the difference between successful predictions that tend to confirm models and successful predictions that, because they lack independence from it, do not tend to confirm the model, Giere supplies us with the model of genetic transfer constructed by Gregor Mendel, and contrasts it with Halley's model of his comet. There is some controversy about the actual history of Mendel's experiments, but according to the "received" view, Mendel first did experiments and then came up with a model to explain the results. To simplify a little by limiting ourselves to a single trait, Mendel noticed that the adult pea plants in his garden were either short or tall and that practically none were anywhere in the large range between. He also noticed that short plants produced short plants whenever crossed with short plants, and he isolated within the tall plants some that always had tall offspring, whether crossed with tall or short. Then he began his breeding experiments. He crossed the tall plants with the short and got a new generation of exclusively tall plants. But the plants of this generation, when bred among themselves, produced both tall and short plants, though the tall outnumbered the short three to one. Each time he performed this experiment, he got the same 3-to-1 ratio. (And not just with the traits tall and short; green and yellow, and smooth and wrinkled, also appeared in a 3-to-1 ratio.) To explain why peas expressed these traits in this ratio when bred as they were, Mendel articulated a model of inheritance involving dominant and recessive traits. Each plant had two "factors," one from each parent. The factor for tallness was dominant, that for shortness recessive. Thus, the members of the new generation each had one tall and one short gene, and since tall is dominant, each plant was tall. And, when the members of this generation were bred among themselves, one quarter had two short genes and three quarters had either one or two tall genes. Mendel's model of inheritance worked well to explain the observations Mendel had made during his experiments. Not only is the model explanatory, one might think, it is also predictive and thereby testable. One can use the model to predict that the offspring of interbred second generation pea plants will again be tall and short in a 3-to-1 ratio. The best evidence that a model accurately represents a system is that the model generates successful predictions of the behaviour of the system. That means, one might think, that the success of this prediction would speak in favour of the model's accuracy. However, the success of this prediction in fact would not go far at all toward vindicating the model. The reason is that it is not improbable for the prediction to be successful though the model is false. One would here be predicting, unlike Halley with his comet, the occurrence of the very phenomena his model initially took as data. It was the observation of these ratios that required an explanation in the first place.

This after-the-fact theory creation to explain observed events is not a bad thing, and is not unusual in science. It is a testament to the ingenuity of scientists. What is at issue here is whether the explanation is the best one or the correct one. Mendel's theories of the genetics of pea plants could not be tested against his original observations because his theory was designed to explain this record. It is relatively easy to theorize about how certain phenomena came to be observed. For example, knowing that the comic adventurer Michael Palin was refused entry into the Amundsen-Scott South Pole Station during his Pole-to-Pole travel adventure, I might suppose that governments are conducting secret experiments on alien spacecraft underneath the ice. This supposition would explain his being turned away. However, I could equally well explain his denied entry by supposing that those in charge of the base wish to discourage further adventure seekers. The way to test whether your explanation is any good, is to see whether it can make any predictions about events not yet known. Mendel's initial observations cannot serve as a test for his theory because they were known in advance of their explanation. The theory was designed to explain them, and thus they cannot be thought of as improbable. It was only when Mendel performed a new "backcross" experiment, where his theory predicted novel results, that his theory was successfully tested. This new experiment involved breeding experiments that he had not already performed and did not know in advance what the results would be. However, using his model of inheritance, he was able to successfully predict that in the third generation of pea plants, the ratio of tall to short plants would be 50/50. Because this result was not known in advance, and was unlikely to be true unless his model was an accurate representation of events, he was able to conclude that his model of inheritance was probably accurate.

I will argue below that the use of historical data to test models of global systems often violates this improbability condition. It is often claimed that models of global climate can be relied upon to accurately portray the future because they have successfully replicated the behaviour of the climate in the past. However, since this historical climate behaviour is used to design climate models in the first place, the successful replication of the past is not improbable at all and does not then constitute a reason for confidence in the predictions of future climate behaviour.

The third condition of a good test is that the prediction be verifiable. This last condition is difficult to fulfill for models of global systems. Many predictions are difficult to verify at all because of the many interference factors; for example, R.C. Balling reports that the measured global warming and cooling trends of the last century are not statistically significant.[3] Even if statistically significant verification is achieved, it might well come too late to be useful. Of course, whether a model is verifiable in the long run is a separate question from concerns about usefulness. However, useful verification is what is important here. Again, a central feature of the problem in question is that the nature of these catastrophes is such that, if they come to pass, then convincing evidence will not be available until it is already too late to do anything about it.

I do not wish to argue that there are *in principle* reasons why any model of a global system cannot fulfill these three conditions. However, I will argue that the models we have right now do not, and it will be very difficult to improve them so that they will fulfill these conditions.

·4·

Two Requirements for Successful Prediction

The three conditions oulined in chapter 3 are meant to serve as criteria for assessing whether a prediction is a good test of a model, theory, or hypothesis. I have pointed out already that for models of global systems that are predicting catastrophe, this direct testing procedure is not useful. We should like to have some method of assessing whether a model is capable of making reliable predictions without having to wait and see if those predictions come to pass.

We can start on this task by assessing a model's ability to satisfy two requirements of successful prediction. In order to predict, we must have an accurate assessment of the present situation and a reliable account of the mechanism that will connect this initial state with a future state.

The initial values we need to insert into the equations of a model include such things as population size and rate of growth, temperatures, diffusion rates, absorption coefficients, and chemical concentrations. (Exactly which data are needed will, of course, depend on the particular model.) These values can be obtained by empirical measurement. Difficulties in measuring, however, are a great source of uncertainty, as will be shown below.

The second requirement for making predictions is that it is necessary to have a reliable description of the mechanism that translates these initial values into future-state values. Even if we are fairly certain of our initial data, and believe that we have accurate descriptions of particular mechanisms or local interactions, we might have reason to be doubtful about the behaviour of the entire, complex model. We may know a good deal about the components and still be unable to generate descriptions of the interactions among these components such that we can make reliable predictions.

There are three issues here. First, there is the question of the *accuracy* of the data supplied to the model. Did we get the initial conditions right? Second, there is the question of the *reliability* of the model. Are the principles of change and development it involves reliable? Is the model complete, or does it leave something important out? Are the elements of the model consistent with each other? Third, supposing we have good answers to the first two questions, there is the question whether the *predictions* the model is said to license are actually licensed by the model. How are the probabilities the model assigns to the predictions arrived at? Are they justified? How is a model tested?

I will argue that global models are weak in all three of these areas. This chapter will focus on the first two criteria. First, the accuracy of measurements of the initial data is suspect, and any inaccuracies might get multiplied. The world's systems are complicated and naturally variable, so measurement of the required variables is hampered by the relatively small sample size of observations and high noise levels.[1] Second, the ability of a model to translate this initial data set reliably into the future is in doubt. The need for simplicity forces models to group together several similar components of the system being modelled and this method of aggregation can be misleading. Problems of reliability are also generated by the fact that several models of the same system are created for various purposes. This raises doubts about the epistemological unity of these different models and the possibility of their combination. The third concern, that of predictive weakness, will be discussed in chapter 6. Global models often turn out to have a tautologous structure that precludes testing even for predictions of the near future. I will present R.H. Peters's argument that some models cannot predict because they are, in fact, tautologies in the sense that they cannot explain anything not already assumed in the model. I will try to show that some current models of global risk fit this category.

REQUIREMENT ONE AND PROBLEMS WITH ACCURACY OF THE INITIAL DATA

The accuracy of the initial data is challenged for reasons of empirical limitations. This inaccuracy raises the question of whether some global systems are chaotic (i.e., sensitively dependent on initial conditions), and hence whether it is possible to make long-term predictions for these systems.

Empirical Problems

The first requirement mentioned above faces the empirical difficulty that the data available on population size, food production, energy reserves, soil erosion, pollution levels, and so on are very sketchy. Accurate measurements are required not only to obtain an idea of the current state of Earth, but also to obtain a record of the past with which to compare the current state. If we do not know how a system has behaved in the past, we cannot know whether any changes are unusual or simply part of a natural variation.

The amount of work required to measure the variables needed for a complex model is so astronomical as to be unrealistic in many cases. Furthermore, these variables are always changing. Sometimes, there are simply no techniques for obtaining accurate measurements of the desired quantities and the estimates can vary enormously depending on who is doing the measuring; for example, that the ocean temperatures needed to study climate change were not recorded before about one hundred years ago. Large areas of the ocean that lay outside of shipping lanes were never even travelled, let alone measured. Nobody ever thought of recording soil moisture. Standard procedures for data collection, such as measuring at the same time of day or at the same ocean depth, have not been adopted until relatively recently.[2] Similarly, Balling notes that changes in calculation methods were not standardized; for example, the mean daily temperature may have been calculated by averaging the daily highs and lows or by averaging twenty-four hourly measurements.[3] He also reminds us that the changing technology of measurement over history can give misleading results; for example, whether sea surface temperatures are taken using a wooden, canvas, or plastic bucket affects the measurement because of different cooling rates. The standard for measurements of air temperature used to be from thermometers on the north walls of buildings, and later from inside weather boxes; it is now from satellites. Most climate measurements are taken near cities and are, therefore, skewed upward because of what is known as the urban heat island effect (where recorded temperature increases might best be attributed to the increased area of pavement rather than to increased carbon dioxide emissions.) Balling thinks this measurement effect might be responsible for up to 25 percent of the global warming observed in the record of the last century.[4] (Although the most recent IPCC report adjusts for urban heat island effects and estimates that the global average surface temperature has increased by about 0.6°C.[5]) In other cases, the sheer quantity of measurements required makes the task almost prohibitive in terms of money and time. Sometimes, for things like the quantity of energy reserves and local population growth rates, the data are simply not available. There have been attempts to adjust the data in order to correct for these sources of error. However, because we have no baseline data for comparison, we do not know if these adjustments are indeed corrections.

Sometimes, the exact nature of the scientific data required is not properly defined. Some concepts – such as biological diversity or the assimilative capacity of the oceans – are too vague or are defined in several ways, and therefore accurate measurement is not an applicable concept because it is not clear what exactly needs measuring. These concepts do not necessarily render the models that contain them useless, however. If the concepts are important and cannot be made precise, we may just have to learn to live with them. Perhaps they may be used as qualitative indicators even if they cannot be quantitatively expressed.

Even when quantitative measurements are routinely gathered, however, the accuracy of these measurements may be inadequate for the task. Even the simplest of systems cannot be described without estimates and small errors. These little inaccuracies in the initial measurements can lead to larger errors when inserted into a model. Because of the recursive nature of the models, additive and multiplicative compounding errors will result. This problem is so great that it has been claimed by some that the predictions of complex models cannot be distinguished from random processes.[6]

This problem can be eased with built-in checks and limits on certain kinds of unwanted numerical growth, but Peters warns that such gerrymandering of models is unscientific because it "constrains the model to act as we believe it should."[7] The parameterizations of global circulation models, for example, are tuned in order to keep a close resemblance between the simulations and observations. These constraints for the purpose of realism, however, force the model to behave in familiar and expected ways. Remember that parameterizations in a model can relate variables without detailing the specific processes involved; for example, it matters little whether the mechanism of distribution of water vapour is represented as some convection process or as large-scale rainfall – so long as the general pattern of global distribution is realistic.[8] Modellers can thus adjust these parameterizations until they achieve desirable results. The Lancaster Centre for the Study of Environmental Change reports, for example, that when faced with a model that made inaccurate predictions of surface temperature, some modellers felt justified in substituting a less realistic parameterization of meridional heat transport in order to attain greater accuracy in the desired area (because the observations of surface temperature were more reliable than the records of heat transport).[9] When a model of the ocean is coupled with an atmosphere model, errors arise due to sensitive dependence on initial conditions. To correct these drifts, the model is tuned by adjusting the amount of heat exchanged between the ocean and atmosphere.[10] Similarly, global climate circulation models are constrained in various ways in order to ensure that the models supply a realistic pattern of seasonal cycles. Another example is provided when models are tuned in order to reproduce complicated patterns in a system, say wind and ocean currents. However, this pattern matching is only possible by constraining the model in ways that are not rigorously justified.[11] Of course, this tuning assumes that we know beforehand exactly what the model is intended to find out. (Although the most recent IPCC report claims that "Some recent models produce satisfactory simulations of current climate without the need for nonphysical adjustments of heat and water fluxes at the ocean-atmosphere interface used in earlier models."[12])

Because of inaccuracy of measurements, lack of standardization in measuring, and lack of baseline data for error correction, the first precondition for successful prediction – that we must have an accurate assessment of the present situation – is only weakly fulfilled by global models.

Chaos

It is sometimes thought that the question of sufficient accuracy might even lead to an insurmountable difficulty in the whole project of modelling the future. According to the mathematical theory of chaos, prediction must inevitably meet fundamental limits for certain kinds of systems. It is argued that systems that are sensitively dependent on initial conditions are intrinsically unpredictable.

A system that is sensitively dependent on initial conditions behaves in ways that are significantly different with only slight changes in these initial conditions. For example, very slight changes in atmospheric pressure or humidity might change the overall behaviour of a weather system. If the world is such a system with respect to the behaviour involved in models of global risk, then we might never be able to determine the future consequences of our present-day actions because we could not make the required measurements with sufficient accuracy.

I will first describe the threat that chaos presents, and then discuss whether the predictions made by models of global systems are undermined by this threat. Some global systems might be chaotic in this way, some might not. Models aim to show whether a system is chaotic or not, and at what level. This is done by seeing whether the models are sensitive to initial conditions or not. If chaos exists only at the micro-level, it might still be possible to make predictions at the macro-level. Despite the inability to forecast detailed weather conditions, it might still be possible to make general predictions of climate patterns. In this case, the possibility of chaos prevents detailed predictions of complete certainty, but models of the future do not pretend to give predictions with complete certainty. They only claim to predict general trends with reasonable assurance. However, if the chaos is at the macro-level, we will be unable to make any long-term predictions with confidence.

In a deterministic universe, if one knew all the laws of a particular system as well as the initial conditions, one could theoretically predict the behaviour of that system. However, this way of thinking is not of much use in predicting the behaviour of turbulent rivers, weather systems, or other complex features of the world – as theoreticians of chaotic phenomena point out using empirical studies. This is because these systems are sensitive to small changes in initial conditions. Furthermore, even very simple systems, like a pendulum, can exhibit chaotic behaviour. A chaotic phenomenon is more than a confused sum of jumbled interferences that is hard to make out. Rather, it is the rapid divergence in behaviour of a system with only slight changes in the initial conditions.

Models of these systems will, therefore, be sensitive to measurement error. The sensitive dependence on initial conditions of a system means that small errors of measurement of the initial conditions can lead to huge differences in predictions made by a model using those measurements. One example of measurement uncertainty is the data missed by meteorological measuring equipment. James Gleick reports that even if we had weather sensors every square foot (horizontally and vertically), meteorologists still could not predict the weather because the errors of extrapolation between points lead quickly to huge errors in large-scale behaviour.[13] Thus, even if a system's behaviour is completely deterministic, our practical knowledge of it will be limited because even very small uncertainties are magnified very quickly. This "butterfly effect," as it is called, results in the apparent paradox that even though the mechanisms of some simple systems might be purely deterministic, these systems can still behave in unpredictable ways.

Thus, in chaotic systems, the two conditions for prediction – an accurate assessment of the present situation and a reliable account of the mechanism that will connect this initial state with a future state – are connected in such a way that prediction is impossible. The measurement of initial conditions is imprecise and this imprecision gets quickly magnified by any attempt to connect these initial conditions to some future state. Nearby points can rapidly diverge in behaviour, and this makes prediction of the resulting chaotic behaviour impossible.

In his book *In the Wake of Chaos*, Stephen H. Kellert advises that we not underestimate the significance of this problem by dismissing it as only a practical problem – a "mere inconvenience." He reasons that, in this case, the distinction between "in practice" and "in theory" blurs. Instead, we should consider the impossibility of prediction to be a transcendental limitation. That is, even though sensitive dependence on initial conditions does not imply nondeterminism, as is sometimes argued, it does make it empirically impossible for models of chaotic systems to make reliable long-range predictions. Thus, it seems self-deceptive to talk as if the chaotic system would be predictable *if only* we had the right instruments, enough time and enough money to specify the initial conditions. The complete accuracy required is impossible to achieve. Kellert claims that when something is always physically impossible, it is virtually the same as theoretically impossible.

Given this state of affairs, one wonders how any model that attempts long-range prediction of a chaotic system could succeed. If chaos theory is correct, and the inescapable precision errors in initial data prevent accurate, quantitative extrapolations of these data beyond a very limited time span, then any quantitative comparison would seem to be pointless.

I think Kellert's thesis is the right way of solving the debate about whether chaos represents an *in principle* obstacle to prediction or merely a practical one. I agree that a "merely practical" obstacle that can never actually be overcome requires a novel philosophical solution. However, although chaos theory has eliminated the possibility of precise prediction, it offers an alternative. The second aspect of chaos theory is the observation that there is sometimes a macroscopic orderliness to chaotic behaviour. Even though we cannot predict the behaviour of all the component parts of a chaotic system, it may be possible to predict macroscopic behaviour without having actually to predict the behaviour of all the constituent parts. This is important for the project of modelling global systems because it means that, so long as our predictions are not too precise, or too far into the future, it might be possible to rely on a model even if the initial data are inaccurate to some degree.

If Earth's climate is sensitively dependent on initial conditions, then we will not have the ability to predict the weather except for relatively short periods of time. However, this does not necessarily prevent us from understanding general patterns. Climate modellers are not interested in precise predictions of weather. Rather, climate modellers attempt to identify *patterns* and use data that have been collected and averaged over long time intervals.

When precise, quantitative predictions are not possible, qualitative predictions can give a general idea of the *type* of behaviour expected and perhaps reveal common features and patterns of all behaviours. Global models can be tested to see whether any particular model is sensitively dependent. For example, as discussed in chapter 1, the "Limits to Growth" model exhibits four basic kinds of behaviour: continuous growth, rapid growth with a slow approach toward a limit, overshoot and oscillation with a series of crises and recoveries, and overshoot of limits and inelastic collapse. After running the model under various initial conditions, the modellers observed that the "overshoot and collapse" behaviour will occur over a wide variation of initial conditions. Other models, too, might display the same general behaviour patterns regardless of small changes in initial conditions.

Some models, however, will not show such consistent behaviour. The IPCC assessment reports that some global circulation models that are predicting global warming are sensitively dependent on initial conditions, especially when two models of interacting components are coupled (e.g., atmosphere/ocean coupling).[14] The nuclear winter models were also found to be sensitive to initial conditions (as reported in chapter 1.) This stability or instability of behaviour over various conditions is a test of reliability for models and will be discussed more fully in chapter 5, in a section about sensitivity testing.

This means that chaos theory does not present an insurmountable difficulty for models of the future. A sensitivity test might be able to indicate the relative strength of various models and reveal whether the system under study is sensitively dependent or not. If it is found that a global model in question does not exhibit any general patterns when tested for sensitivity, then any long-range prediction will be impossible. On the other hand, some models of global systems will pass the sensitivity test, and in these cases long-range prediction becomes meaningful.

REQUIREMENT TWO AND PROBLEMS WITH OUR UNDERSTANDING OF MODEL RELIABILITY

The second requirement for making predictions is that it is necessary to have a reliable description of the mechanisms that translate a set of initial data into a prediction. The idea that we can know the reliability of a model is challenged by two epistemological problems. First, it is doubtful that a family of models created for specific purposes can be united in a way that will give reliable predictions. Second, there is no way to know, short of trial and error, whether a model of a global system is overgeneralized or overspecialized.

Requirement Two: Epistemological Unity

Because of the complexity of the systems being modelled, modellers adopt the methodological convenience of representing the world as a concatenation of models. For purposes of manageability, several theoretical models are usually produced – each with a specific concern. One global climate model might specialize in clouds, for instance, and another in oceans. This multiplicity is routine and is why Giere adopts the "family of models" approach outlined in chapter 2. Similarly, Nancy Cartwright claims that "no single model serves all purposes best."[15] However, the ultimate goal is to make predictions about the entire system – and about this none of the specific models can yield precise predictions. It is often assumed that the result of a complex interaction can be determined by combining the results of various models and resolving the outcome – rather as in vector addition. However, we cannot tell what will happen when we combine models because, as Cartwright has argued, our causal theories have been created for "separate domains."[16] The combination of model-results is, then, not just a matter of difficult calculations and rough approximations. There is an epistemological objection to this procedure because if the various models are constructed for different domains, then their incommensurability might make their combination pointless.

Ian Hacking also argues that there is a lack of epistemological unity between models. He says that science works by manipulating "limited models of little bits of manipulable materials" and that these models are inconsistent with one another.[17] This is why, Hacking argues, new technologies often produce undesirable "interference effects." While the behaviour of individual component systems might be well understood, their interactions can produce unexpected results. As an example, Hacking describes the pollution control devices of coal-fired power plants, which involve the turbulent flow of fly ash through electrostatically charged baffles. Hacking claims that while the behaviour of fly ash under either turbulence or electrostatic charge is well understood, the interaction of these two forces is not merely an average or compromise between the two: "The behavior of fly ash under conditions of turbulence and electric charge is altogether unlike what understanding of either or both sets of laws would lead one to believe."[18]

This argument is significant because members of a family of models are often linked together. One global circulation model, for example, emphasizes the behaviour of oceans, while another specializes in clouds. If Cartwright and Hacking are right in claiming that we cannot simply assume that the results of these different models can be added together, then we have a serious lack of completeness because the interaction between these components is fundamental to the behaviour of the whole system.

Aggregation

We have seen that there are quantitative problems in obtaining an accurate assessment of the initial conditions and extrapolating, via representations of mechanisms, from this assessment to a future state. In addition to the quantitative uncertainty produced by difficulties in accuracy of measurement, there are *qualitative* problems with models that are a function of the model's structure. The method of aggregation creates epistemological problems because it is not known whether the model is too detailed or overspecialized.

Aggregation is a structural method that models use to simplify matters by grouping together several different quantities and assuming that they can all be lumped together and treated in the same way; for example, composite parameters are often used for things such as pollution, natural resources, and population. Global mean average temperatures are calculated instead of regional temperatures or day and night temperatures. Some greenhouse models have attempted some regional variations, but it is admitted that models are, in general, not detailed enough to provide reliable predictions for specific regions.[19] Similarly, the early nuclear winter models represented the world as simple, uniform, and undifferentiated, while ignoring seasonal and geographical variations.

However, ignoring the actual differentiations in these parameters (between developed and developing nations, rich and poor, local geography, culture, consumption levels, development costs, for example) might make a significant difference to the structure of the problem; for example, the "Limits to Growth" model has been criticized for treating all pollutants as if they all behaved in the same way. If so, that assumption ignores the fact that some pollutants have negative feedback loops while others have positive feedback loops. Similarly, aggregation of all pollutants misses the fact that air pollution is localized mostly in cities and stems mostly from cars. Thus, this aggregation might lead to misleading conclusions about the limits of an entire system when, in actual fact, there might be only localized limits and regional problems.[20]

This objection against aggregation is more than a mere insistence on finer detailed conclusions. It might be that the very structure of some of the world's interactions are misrepresented as global in nature when, in reality, they are not.[21] This can then lead to the creation of spurious problems that trap us into being overly concerned about nonexistent problems while ignoring real ones. R. McCutcheon writes that "by aggregating ... and assuming they behave in some composite way, attention is drawn away from what are urgent and still soluble problems and diverted into speculation upon an imaginary race against time between 'life' and 'global asphyxiation.'"[22]

In other words, we end up erroneously focusing on population growth, say, or development in general, when perhaps we should be focusing on the *distribution* and type of growth.[23] If the problems of ozone depletion, or population explosion, start at a local level and are regionalized – rather than occurring all at once in a global collapse – then, perhaps, the final consequences will not be catastrophic. (I mean "not catastrophic" in the sense of global and irreversible. Obviously, the results could still be tragic.) If so, this changes the whole decision problem because the thesis is focused on the idea that decisions must be made to avert catastrophe of a kind that gives no warning, no time to react. Similarly, the concept of a "threshold" in the nuclear winter models turned out to be "an artifact of a simplified model."[24] Jettisoning this concept changed the entire nature of policy implications. Aggregation also masks important features of global warming. Balling holds that the predicted warming will most likely be manifested in a reduction in the diurnal temperature range (the difference between maximum and minimum daily temperatures), with the warming taking place mostly at night and mostly in winter.[25] This important, and perhaps noncatastrophic nature of the warming is masked by aggregate reports of global temperature averages. Of course, even a global warming that is felt mostly at night might still be disastrous, as the scope of water-borne and mosquito-borne diseases would likely increase.[26] However, the problem of global averaging cannot differentiate between the two problems, and each deserves its own risk assessment.

To group and average in this way is to commit the fallacy of composition. However, generality is needed if the model is to capture the big picture and not get bogged down in details – simplification is not always a negative thing. For some purposes, a general representation is all that is needed to identify consistent behaviour patterns. In fact, some modelling techniques are structurally incapable of giving a more differentiated picture. If general representation is the aim of a model – rather than precise prediction – then failure to give a differentiated picture is not really a fault. Global circulation models, for example, while not able to represent regional climates, are still considered useful if they are able to provide global equilibrium temperatures. The job of a modeller is to start from general equations and then to add enough detail to examine a particular issue. The appropriate level of aggregation will be a function of the effect one is trying to model.

While aggregation is admittedly a problem, unnecessary or *over*differentiation can also increase uncertainty. A faithful representation of national boundaries and the various internal social policies might obscure the larger picture that might have to do with global, physical limits and be little affected by local differences. Another potential problem with attentiveness to detail is the overzealous tracking of "ripple effects" (tracing a causal chain to extreme lengths) that could lead to a false confidence in the precision of the results. As a fictional example, one episode of the cartoon *The Simpsons* portrayed a model of an imminent meteor strike that had everyone convinced that there was no threat except from a fragment that was expected to land on Moe's bar. A real-life example of false precision might be a model indicating the direction of winds coming off the ocean when normally they blow out from the land.

Models must walk the line between overgeneralization and overspecification. However, there is no way in principle to know whether a model of a global system is overgeneralized or overspecialized or, indeed, walking the line between. The problem is that we do not know whether the category divisions in a particular model represent important categories of mechanism in the actual world. We do not know, for example, if "pollution" is a more or less useful way of categorizing phenomena for use in a model than a more detailed division into various kinds of pollution.

The problem is that of carving up and categorizing the world into classes of objects that will facilitate better explanations. Whether any given categorization accurately reflects the way the world is actually cut up is known as the problem of natural kinds. An analogy here might be useful in explaining the concept of natural kinds. In *Empire of Signs*, Roland Barthes describes how the use of chopsticks is very unlike the use of knives and forks.[27] While knives and forks can cut up a food into any arbitrary shape, chopsticks find natural divisions that gently divide the food in a non-arbitrary way – where the peel meets the fruit, for example, or the sections of an orange. Similarly, scientists aim to capture real divisions rather than arbitrary ones. We suppose that we have discovered these divisions when the class defined by the division has a meaningful role in models, explanations, or physical laws. So, for example, the class of "spider plants" is a natural kind because there are all sorts of biological laws and explanations about spider plant nutrition, reproduction, and appearance. On the other hand, the class of houseplants isn't a natural kind. Knowing that a plant is a houseplant does not enable one to say much at all about it. For another example, consider fire, rust, and the stars. It was once thought that fire and the stars belonged to the same class of objects, while rust belonged in another category. However, our current division of the world's objects groups together fire and rust as processes of oxidation, while stars are in a class related to nuclear fusion. The current division gives scientists more prediction and control.

In the case of modelling, we want to capture the divisions that will lead to interesting predictions, and the worry is that some categories like "pollution" might not lead to useful predictions. There are many kinds of pollution that act in many different ways rather than as a general class. On the other hand, we do not want to be overly detailed either. For cases of "everyday" models like the traffic model example of chapter 2, the modeller needs simply to hypothesize about which categories are important to differentiate and which are not and then test the hypotheses by trial and error. It might turn out that we can treat cars as sets of waves and get good predictive power, or it might turn out that better predictions are achieved by individuating the cars. However, as has been noted before, for models of global systems predicting catastrophe the method of trial and error is not applicable. In the unique cases of global risk that we are considering, modellers do not have the luxury of waiting to see what happens and whether the category divisions used are the best ones. Therefore, modellers do not know whether the categories specified in such a model are too general or too specific.

CONCLUSION

We have seen in this chapter that we cannot be certain whether some particular model satisfies the preconditions for successful prediction. Problems with inaccuracy and lack of standardization in measurement mean that we have weak baseline information for comparison when making judgements of change. Our understanding of the reliability of models is not especially good. First, we cannot unify models because each conglomeration is essentially a new model. Second, we do not know if we have identified the proper level of specification needed to identify the categories important in the mechanisms involved. Even construed as charitably as possible, these issues are not going to be settled short of obtaining the predictive adequacy that we do not wish to wait for in cases where models of global systems predict catastrophe.

· 5 ·

Assessing Models of Global Systems: Indirect Tests

I have pointed out that standard ways of evaluating scientific models by trial and error do not meet the requirements for distinguishing accurate global models from the not so accurate ones when predictions of catastrophe are involved – because we need to make the distinction without having to see whether predictions of catastrophe come to pass. Given that models of global systems cannot be tested by direct methods of prediction and trial and error, and that we are not sure whether some particular model satisfies the two requirements for successful prediction, how are we to defend predictions based on models of global systems that cannot be directly tested?

Science has resources for evaluating theories even when the best, most reliable method – the direct testing of implied empirical consequences – is unavailable or insufficient. What might be called "indirect" methods of determining the strength of a model must suffice, or we will forever be precluded from any long-term planning. Historical agreement, predictions of the near future, sensitivity analysis, and analogy have at various times been used to strengthen belief in the accuracy of a model – even though no prediction has been verified.

HISTORICAL AGREEMENT AND PREDICTIONS OF THE NEAR FUTURE

Modellers often appeal to the consistency of a model with historical measurements as a measure of its accuracy. If we can make the model agree with the past and accurately predict the near future, then this indirect verification should increase the strength of our belief in its accuracy. Such historical agreement speaks in a model's favour – not as an absolute criterion of epistemic goodness, but as a baseline criterion (successful historical agreement makes the model worth looking at) and, more important, as a comparative criterion (a model better at representing past events than a competing model has something in its favour over that competing model).

Historical agreement is about two things. Not only is the extent of agreement important (lots is better than little), but so is the nature of the agreement (surprising fits are better than routine, expected fits). If, for example, a model interprets past data in a way that explains previously unsolved anomalies, or reproduces (i.e., simulates) complex behaviour not used in the construction of the model, then it passes an independent test even though the data are already recorded. The NASA model of the greenhouse effect reproduces various and detailed climate changes – rather than just a global average temperature increase – and when a simulation of the past hundred years was run, it successfully managed to reproduce the appropriate seasonal trends.[1] Other models are said to produce a reasonable simulation of more infrequent events such as the El Niño Southern Oscillation (a semi-regular climatic phenomenon of interaction between the atmosphere and the tropical Pacific Ocean).[2] Other indicators of accurate simulation include the reproduction of sea-level pressure patterns, zonal winds, regional eddies, precipitation, soil moisture, snow and ice cover, and monsoons.[3] One model, developed by the Hadley Centre for Climate Prediction, claims to be the first model to successfully represent the historical pattern of climate fluctuations since accurate record-keeping first started in 1860, accurately reflecting temperature increases and plateaus following the industrial revolution and the second world war.[4]

Unfortunately, the 1990 IPCC assessment reports that, when it comes to smaller scale, regional simulations, there are "significant errors in all models" of global climate.[5] Since then, modellers have attempted to improve regional simulations. However, even if models were more successful in this task, there would be two problems with using historical agreement as an epistemic criterion.

First, there is a question as to how much historical agreement is required. In one actual example from the recent past – the "Limits to Growth" model (called "World3") – this agreement is not obtained except for the relatively short period of the last seventy years.[6] Or perhaps a model might correlate well for one historical period but not for another. In fact, an earlier version of the World3 model, historically correct for this century, was run starting with data for 1850, and it predicted a doomsday-type collapse for 1970![7] As already mentioned, this criterion is a comparative one, not an absolute one; but such limited successes can only justify limited faith. Many models also assume the existence of thresholds below which no significant behaviour deviations will occur. If these thresholds are not exceeded in the historical test period, then again the test is inadequate.[8]

Second, even if the model is tuned so that there is no disagreement between its representations of historical events and historical observation, this consistency alone does not justify a belief in that model. If the model is *designed* with these historical data in mind, then it is not surprising that agreement is obtained, and this can hardly constitute an independent test. Giere's improbability condition, explained in chapter 3, is meant to rule out predictions that are expected. Remember, he notes, that "to have a good test ... there must be a verifiable prediction that is implied by the hypothesis and is highly improbable given everything else known at the time."[9] Agreement with the very evidence used in constructing the model cannot be considered improbable. It is merely a restatement of the evidence. To then go back and use these same observations as evidence for the truth of the hypothesis is circular. global climate circulation models, for example, have been accused of circularity because the agreement between observations and model simulations is often achieved by choosing the model parameterizations specifically in order to create the match.[10] Parameterizations are manipulated until the model reproduces expected seasonal cycles, expected high and low pressure troughs, and the like. Balling reports how some models, in order to fix anomalies in their historical simulations, have been adjusted by increasing the value of solar radiation to 7 percent above its known value, or lowering the reflectivity of clouds by 10 percent below its known value.[11] This "tuning" of models to match known observations violates Giere's improbability condition. So even though the greenhouse models may accurately recreate past climates, recreating seasonal variations where expected, etc., this is insufficient to differentiate them from models that manage to achieve the same historical agreement but that predict no catastrophic warming. While organization of past events into a model can help provide an explanation, it does not provide the good predictions – predictions that satisfy Giere's three criteria – that are necessary to test the model.

A similar indirect test is the use of predictions of the near future as a limited form of verification. However, most successful predictions of the near future are ruled out by the same improbability condition just mentioned above. Predictions of the near future rarely give strong corroboration because they are usually predicted by almost all competing models. The "Limits to Growth" model, for example, predicts "overshoot and collapse" of population growth, while others predict a slow, gradual approach to a relatively high level. Now, predictions about what will happen, say, ten years from now would reinforce neither alternative because even if these predictions come to pass they would likely be consistent with several models.

Similarly, the amount of global warming measured to date is compatible with both the greenhouse models and natural variability. The evidence for global warming comes from many sources. Analyses of samples of ancient air from glacial ice core samples reveal a steady increase in the concentration of greenhouse gases, gases shown in laboratories to have warming effects. Meteorological records show a warming trend (the 1990s was the warmest decade on record). Canadian lakes have an increasingly long ice-free season. The area of permafrost in the MacKenzie River delta region is decreasing. Some of the world's mountain glaciers are retreating. Sea levels are slowly rising. All this evidence is consistent with predictions made by greenhouse models. However, the thesis that greenhouse gas-induced global warming is being dangerously increased by anthropogenic emissions is not supported, because the predictions above fail to meet Giere's criteria for a good test. The above events are not improbable because other explanations for these observations remain possible and plausible. Although the fact that greenhouse gases are accumulating is well established, it is impossible to prove that these gases are increasing global warming.

Even if the presently observed correlation between rising temperatures and rising greenhouse gas concentrations were not disputed, this correlation is not itself enough to warrant the conclusion that the increased concentrations are causing the temperature rise. Continued observations of such a correlation for the next century would be enough – because such a correlation would be highly improbable unless the theoretical models predicting this correlation were similar in important ways to the real world. Once again, however, such observations will come too late to be useful.

The ambiguous success of predictions of the near future provides an explanation for why some scientists state their findings ambiguously; for example, Gates claims that recent temperature increases "are *not incon-*

sistent with the general trend of (the greenhouse effect) models" (my emphasis), suggesting that the temperature evidence supports the increased greenhouse hypotheses even though it is also consistent with natural variation.[12] Others, however, are not as cautious. H.W. Bernard writes that "scientists say the probability of such changes occurring merely by chance are only one in 100."[13] I think that, given the lack of novel and improbable predictions of the near future generated by the greenhouse models, these latter scientists are going beyond what they are justified to say.

Significant and rapid changes are not predicted until approximately the year 2020,[14] and only then would the difference be great enough to be convincing – if it comes to pass. Until then, scientists will attempt to elaborate the models so that differences in more fine-grained predictions emerge. However, at present, fine-grained differences, which can be accurately measured, are not available. In the meantime, there are experiments or measurements we could, in principle, make to gain evidence about the possible feedback mechanisms. It has already been established, for instance, that the surface ocean currents are incapable of absorbing sufficient carbon dioxide to slow global warming.[15] Tests could be devised to see whether the deep ocean currents are currently absorbing greenhouse gases through interaction at the boundaries of these two types of ocean currents. Such tests might rule out or confirm the possibility of some of the suggested consequences of continued greenhouse gas emissions. This information will serve to make the models more elaborate – but we are in a position of having to decide what to do about global warming before waiting for this kind of information, which might or might not improve the predictive power of the models.

Thus, at best, historical agreement and near future prediction can narrow the field of competing models by ruling out those that fail to meet the minimum condition of matching up the behaviour predicted by the model with past courses of events. Any match could not perform more than this negative role because it would be too easily achieved. It would be too easily achieved because historical data are used to construct the model in the first place, and near future predictions are not sufficiently detailed to be differentiating.

SENSITIVITY ANALYSIS

Whether a model is robust in the face of variations in assumptions about the initial data is another indirect method of testing available to modellers. By "robust," I mean being consistent in behaviour and predictions over a wide range of assumptions. A model can be tested to see whether the behaviour it predicts remains generally the same, even if we have different assumptions about particular mechanisms or relations within the model. If it does, then we can have confidence in the reliability of the model's predictions, even though we lack precise initial data. The stability of the predictions indicates that the predictions are trustworthy and are not dependent on unusual circumstances, erroneous data, or false assumptions. Somewhere in this wide set of assumptions and adjustments, it is thought, we must have included the correct data set.

In addition to supplying statistical ranges for predictions, this test can give qualitative information. By revealing where the model is stable and where it is sensitive, sensitivity analysis can identify where we need to make more precise measurements or where we can influence the behaviour of the real-world system. Modellers believe that stability reveals basic structural patterns of the system under study. For example, the main idea behind the "Limits To Growth" model was that the global system cannot be understood as simply the sum of its individual parts. Therefore, this model attempts to analyze the general tendencies of the global system to behave in certain ways *without* having to rely on precise information about all the details of the individual parts. With this idea in mind, the modellers analyzed many different possible futures. They wanted to see, for example, what would happen with their model if we assumed new discoveries of vast resources, or new pollution controls, or incredible advances in technology. What the results of this sensitivity analysis revealed was that the general behaviour pattern of the model was always similar, except in the cases when population growth was checked. The pattern revealed that, for a wide range of assumptions, a global collapse was imminent. This collapse might come a little later, or for a different reason (economic collapse rather than resource depletion), but the trends were always the same. The only way to escape the collapse mode on their model was through constraints on further population growth. Of course, these conclusions have not gone unchallenged. Despite the authors' claims that the model represented "basic behavior modes ... so fundamental and general that we do not expect our broad conclusions to be substantially altered by further revisions,"[16] others have claimed that their own testing of the model revealed extreme sensitivity to changes in input and that the inputs had wide error margins.[17]

Models for greenhouse warming similarly ignore the details of local temperature changes and attempt to represent more consistent global patterns under varying conditions. Some claim that the warming models are robust because their representations of warming are consistent despite changes in assumptions regarding rates of gas concentration increase, or cloud formation. For example, Bernard writes that "Computer models have become more sophisticated, but the answers they are giving haven't changed much. The current generally accepted range for greenhouse warming resulting from doubled carbon dioxide is 1.5 to 4.5°C."[18]

In contrast, the nuclear winter model was not robust. This model was extremely sensitive to minor changes in the initial data estimates. The results changed dramatically depending on assumptions made about the amount of combustible material in cities, the amount of soot produced (as opposed to dust, which reflects less light), whether the nuclear exchange took place in the summer or winter, and the "spottiness" of the cloud formation.[19] In his "memoir" on the subject, Rothman claims that he "could pick assumptions and data which are as valid as the TTAPS [the name of the model] choices, and nuclear winter would essentially vanish."[20]

The fact that nuclear winter might not ensue should a nuclear exchange take place in, say, autumn does not bankrupt the theory that we are at risk of having a nuclear winter, as, for one thing, there is still the summer to consider! The fact that a model specifies conditions under which some catastrophe would occur, conditions that do not always hold, does not make the model any less respectable. However, a conclusion that is dependent on many questionable assumptions is not as strong as one that is independent of these assumptions or has a wider margin of error. The nuclear winter hypothesis has not been falsified by its lack of robustness, but confidence in its predictions has been considerably weakened by its dependence on questionable assumptions about initial parameters.

Unfortunately, such sensitivity analysis reveals that most global models are not robust. The complexities created by complicated feedback processes, mentioned earlier, once again cause problems. Very small changes in assumptions about the nature of these feedback loops can completely reverse a behavioural trend.[21]

Furthermore, the confidence gained because of a model's stability over a range of datum inputs might be illusory. Perhaps it should not be surprising that once we have adopted a certain model structure – feedback loops, system dynamics, large grid climate measurements – the behaviour remains basically unchanged with respect to changes in precise details. Balling writes that "early GCMs [global circulation models] ... were bent on producing a warm planet no matter what."[22] Sensitivity analyses can identify stability or sensitivity to changes in data, but the stability might be due to structural features of the model, and we are uncertain whether to believe in the structure of the model. Sensitivity analysis will not change this uncertainty.

The biggest drawback to sensitivity analyses, however, is a practical one. Systems models are so complicated that only the most expensive supercomputers are capable of manipulating the data. Not only are these computers scarce (and therefore expensive), but they are still relatively slow at producing results. Shackley et al. report that general circulation models require "massive computing and personnel resources," which "precludes the opportunity of performing extensive sensitivity and uncertainty analysis."[23] To run a climate simulation using the most sophisticated, coupled ocean-climate models, for example, might take from three to nine months. This is impractically long because a proper sensitivity analysis involves extensive trial runs that explore the limits of uncertainty of multiple variables. Such an analysis is not simply a process of changing one parameter at a time. At least one hundred trial runs, randomly adjusting various model parameters within the accepted range of uncertainty, could take many years with current computer technology.

ANALOGUES

The models of global systems with which this book is concerned deal with things with which we have no direct experience – nuclear winter, global warming. We do, however, have experiences with volcanic eruptions lowering the global temperature, if not nuclear winter; with extended drought, if not global warming; with serious famine crises, if not collapse of the world's entire agricultural production. We can use these experiences as analogues to evaluate, indirectly, models of the future. If we know from experience that, for example, increases in climate temperature are associated with drought and storm activity, then we can infer that the same will be true for future cases of temperature increase. If the models predict these same associations, then we have reason to think that it is a good representation of what will actually happen.

In his book *Global Warming Unchecked*, Bernard emphasizes this method of indirect testing again and again. His major analogy is between the drought-stricken decade of the 1930s in North America and the near future of predicted global warming. He notes similar patterns of drought in the US between then and now, and, on the basis of this analogy, he predicts more frequent and severe hurricanes, drought in the Midwest, and disastrous crop failures in the near future. These inferences from analogy are the same as those predicted by models of global warming. Bernard writes, "Climatic analogs back up some of these climate model predictions [predictions of increased warming] and where there appears to be an agreement between what the computer models and analogs are telling us, we can feel much more certain that the climatic changes foreseen really will happen."[24] (Not all climate models, however, predict a change in frequency or intensity of tropical storms. In fact, what are called mid-latitude storms are predicted to decrease in severity.[25])

The dust storms that could be caused by nuclear war, along with their expected cooling effects, have analogies in volcanic eruptions such as the 1991 Mt Pinatubo eruption, in Martian dust storms, and large forest fires such as the Alberta smoke pall of 1950.[26] Comparisons between these events and climatic models for nuclear winter were instrumental in both the formation and the later weakening of the nuclear winter hypothesis.

Other analogies can be drawn with ancient historical climates (paleoclimates). Analyses of Antarctic glacial ice cores have revealed a correlation between global temperature and concentration of carbon dioxide and methane.[27] We can also make rough estimates about precipitation levels of various paleoclimates.[28] These historical studies indicate the range of climate variation in the past, and thus suggest what is possible in the future.

Bernard feels that the modelling of the climates of other planets also provides confirming analogues.[29] The warm climate of Venus and the frigid climate of Mars have both been represented using modelling techniques of the kind used for Earth's climate. The agreement between the observed climate on these planets and the climate predicted by atmospheric models supports the belief that models of Earth's climate, which use the same principles, are also reliable. If climate models can represent the climates of other planets, and they use the same principles and share many important features with models of Earth's climate, then we have reason to think that the models will be similarly accurate. The weakness of this test, as might be expected, is that the disanalogies often outweigh the analogies (as will be shown below in a specific example).

AN APPLICATION OF INDIRECT TESTING

The above methods of indirect testing are meant to provide a network of support for a model. Even though it might not be possible to test a method directly, if several indirect tests are in agreement, then faith in the model is strengthened. The problem with this approach is that the support it provides is not detailed enough. For example, as mentioned previously, there is actually very little disagreement among atmospheric scientists about the existence of global warming: the debate is about the quantitative details. The *amount* of temperature increase matters a great deal, as do the rapidity of the warming and the nature of the effects. None of these important matters could be properly treated without a more rigorous scientific model capable of making quantitative predictions. What is to be feared about global warming is the unknown – it is not the actual predicted temperature increase but the *uncertain* effects of such a potentially drastic change. We are uncertain whether global warming will totally disrupt global precipitation patterns with disastrous effects, or whether it will merely shift the climate in a way that is relatively harmless (e.g., the nights get warmer, but the days do not get much hotter). We do not know whether Earth has a self-regulating feedback loop that will operate to readjust and maintain the global average temperature. (However, we know that self-adjustment to maintain a constant global temperature has not always operated, since there have been global temperature shifts before.)

To see why this lack of detail is important, we need to examine an actual case of supposed model confirmation by these indirect methods. We will see that, when we try to test the model using the above indirect methods, the model usually comes up short here and there. This failure is then explained by ad hoc assumptions intended to save the model. In a special issue of *Scientific American* called "Managing Planet Earth," Stephen Schneider examines models of global warming and admits that the models can give neither reliable nor specific regional forecasts. Nevertheless, he goes on to argue that there are grounds for confidence in the general forecasts.[30] This confidence is based on several of the methods of indirect testing described above. A model as a whole "can be verified by checking its ability to reproduce the seasonal cycle."[31] Analogues are also used: "Simulations of past climates – the ice ages or the Mesozoic hot house – serve as a good check on the long-term accuracy of climate models."[32] Analogies to the greenhouse effect on the climate of Venus are also made. Historical agreement is also considered:

"When a climate model is run for an atmosphere with the composition of 100 years ago and then run again for the historical 25 percent increase in carbon dioxide and doubling in methane ... most models yield a ... warming of at least a degree."[33] The observed warming to date is only about 0.5°C, yet Schneider feels that this is a confirmation nonetheless. He claims that "The discrepancy between the predicted warming and what has been seen so far keeps most climatologists from saying with great certainty (99 percent confidence, say) that the greenhouse warming has already taken hold. Yet the discrepancy is small enough, the models are well enough validated and other evidence of greenhouse gas effects on climate is strong enough, so that most of us believe that the increases in average surface temperature predicted by the models for the next 50 years or so are probably valid within a rough factor of two. (By 'probably' I mean it is a better-than-even-bet.)"[34]

I think this confidence is not justified on the grounds of these indirect tests. A discrepancy of 0.5°C might not seem like much, but we are talking about a theory where small differences count for a lot (especially since, in this case, this represents a 50 percent error); 0.5°C is within the range of natural variability and the range of measurement error. Balling reports that when the temperature record is corrected for lack of standardization, the urban heat island effect, and other effects, then the observed warming of 0.44°C disappears. Even if Schneider is using the corrected temperature record, the observed increase of 0.5°C is clearly within the range of measurement error. Similarly, agreement within "a rough factor of two" is a significant error margin when we are trying to determine whether the warming is outside normal variations. Schneider claims that the models are "well enough validated," but this is only because, when there is disagreement between the models and empirical observations, ad hoc hypotheses can be created to save them. To explain the discrepancy of 0.5°C, Schneider supposes that we might have underestimated today's warming because of an inaccurate or incomplete historical record, or overestimated the increase in greenhouse gases, or that some unknown factor is delaying the warming, or that the model is too sensitive for small increases of carbon dioxide, or that the heat capacity of the oceans is larger than thought, or that the sun's output has declined slightly, or that recent volcanic eruptions have resulted in cooling effects that have compensated for the warming. Similarly, the IPCC assessment suggests adjustments that can be made to bring a model into line with observations. For example, it suggests that the capacity of the oceans to absorb carbon dioxide might have been underestimated, or that there is some, as yet unidentified, terrestrial process of carbon dioxide absorption, or that estimates of carbon dioxide production have been too high.[35] This kind of ad hoc explanation allows scientists to save whatever model they are committed to.

Balling has also argued that models of global warming fail to pass indirect tests. The short-term predictions are wrong, and the inferences from analogy are outweighed by disanalogy. The concentration of carbon dioxide has already increased by 40 percent over the last century. (This estimate is achieved by converting all greenhouse gases, with respect to their warming effect, into an equivalent concentration of carbon dioxide.) If the greenhouse models are correct, Balling infers, the increase should already have produced observable climate changes. For such an increase, most models predict a temperature increase greater than 1°C, yet the best data available reveal an increase of less than 0.5°C. He claims that there is no evidence of a causal link between the observed increase in greenhouse gases and any significant temperature increase.[36] Even if the warming is real, the timing of the warming is not consistent with model predictions. Balling reports that most of the measured warming occurred before the Second World War and, therefore, occurred before the bulk of the greenhouse gases were emitted.[37] (The most recent IPCC report notes that there have been two periods of significant warming. The first occurred before 1945 and likely resulted from natural causes rather than anthropogenic emissions. The second warming period, however, from 1976 to the present, is considered to be likely caused by human activity.[38])

The analogy between global warming and the dust bowl of the 1930s also fails because records show that soil moisture in the United States has actually been increasing, and there is no evidence, according to Balling, that the frequency of extremely high temperatures will increase.[39]

Some observed climate changes are consistent with model predictions for the 40 percent increase in carbon dioxide equivalent. The stratosphere has cooled, precipitation levels and cloud cover have increased, and sea levels have risen slightly. However, these observations are consistent with other explanations and, therefore, do not very strongly support the thesis of catastrophic warming. Balling also claims that even if the observed climate changes are caused by increased emissions of carbon dioxide, the magnitude of these changes is small. If a 40 percent increase in greenhouse gas concentration produces such small changes, then this suggests, he concludes, that no catastrophe is forthcoming.[40]

·6·

Dilemmas and Defences

POSSIBILITIES, NOT PROBABILITIES

The examples in chapter 5 show that indirect tests of global models should not generate confidence in their predictions. These arguments were not addressed merely to the general epistemological problems that attend all belief. All beliefs are uncertain in a general, skeptical sense. The activity of projecting trends into the future, for example, involves deep Humean problems about induction. However, if my criticisms of models of the future were all of this strong and far-reaching kind, they would be too recondite to be relevant to the particular issues I am discussing. My concern is not with skepticism and anti-skepticism, but with epistemic difficulties peculiar to the problem of assessing a model's reliability without the benefit of direct testing through trial and error.

The epistemic problem of particular concern for scientific models is that the predictions a model is said to license might be mere logical implications that can give information about what is possible but none about what is most likely to happen. As an example of this problem, consider an equation that forecasts population growth under "ideal" conditions. One such equation – known as "the logistic" – describes an exponential growth that is checked by some environmental limit and, thereby, levels off at some plateau. However, sigmoid-shaped population curves often fail to represent actual patterns of population growth (for many reasons that can be determined on a case-by-case basis after the fact). Thus, the model of ideal population growth provides only a general guideline. Now, as R.H. Peters points out, such general idealizations are not really of much worth because they can only indicate that, when there is no agreement between the ideal case and the actual case, then some other explanation is required.[1]

This reasoning can be applied to global models. When we attempt to understand some aspect of the world through the use of such models, we expect that either the model will be applicable to a situation of interest, or it will not. If it is applicable for the case in question, it succeeds in predicting and explaining observations, and we are satisfied that the model is a good representation of what is actually happening. If, however, observations do not fit with the model's predictions, we can still be satisfied with the model because we can then explain why there are deviations from the ideal in this case. However, we never know beforehand whether a model is applicable to a situation or not. When used this way as an ideal, a model can never be wrong and, therefore, can never be falsified. The only thing it can reveal is that, if the system is not behaving ideally, then some further explanation is required – but it cannot inform the content of that explanation. It is not the *model* that supplies the required additional explanations, but rather the researcher supplying explanations after the fact. Peters's point is that any model that does not specify in advance when it will apply will not be useful as a predictor. It is Peters's worry that too many ecological models are such logical constructions with no empirical application. He insists that these toothless cogs with their tautologous predictions cannot be applied productively to the real world.

K.S. Shrader-Frechette and E.D. McCoy make a similar attack on complex models in their book *Method in Ecology.* They note, for example, that the International Biological Programme, which modelled ecosystems as an interaction between a set of ideal systems – mass flow, energy and nutrient cycles, and the like – "could provide no precise theories having predictive power ... [and was] unable to provide precise predictions that could be confirmed."[2] These models were intellectually appealing, but lacked predictive power.

As I suggested in chapter 5, the global models of the future are easy targets for this kind of objection. Many conceptually appealing models make predictions to which modellers need not be committed. For example, while the existence of a measurable global warming trend would be seen by many as support for the model of the greenhouse effect, the absence of this warming is not capable of disproving the model. Differences between predicted global temperatures and actual global temperatures are easily explained away after the fact. Suppositions about how climate *change* has occurred without the predicted warming can be easily generated. In short, it is too easy to accommodate a wide variety of observations with explanations introduced for no reason other than for this accommodation.

These ad hoc adjustments are not unusual in science. W.V. Quine has argued that theories can always be adjusted in such a way that they will accommodate newly observed phenomena. In "Two Dogmas of Empiricism,"[3] Quine argues that any individual part of a theory may be revised without changing the empirical content of the whole. He points out that our language, beliefs, attitudes, and observations are all interconnected, and a decision to fix one part of this network partially determines the other aspects. The implication of this is that, logically speaking, any theory can be "saved," even in the face of apparently disconfirming evidence.

For example, in the eighteenth century, when Priestly and Lavoisier performed experiments to study a supposed substance known as "phlogiston," they each interpreted the results of these experiments in different ways by making different adjustments to their belief network. Phlogiston was a mysterious substance that was postulated in order to explain observations about combustion. When a material like wood is burned, its volume decreases. This could be explained by supposing that the phlogiston had been ejected from the wood. Priestly and Lavoisier each set out to explore this phenomenon and performed experiments involving the combustion of mercury in an enclosed atmosphere. According to the theory of phlogiston, the combustion should have two effects. First, the mass of the mercury should decrease as the phlogiston escapes during combustion, and, second, this escaping phlogiston should increase the volume of air in the enclosure. This was the opposite of what was actually observed. Now Lavoisier took the results of this experiment to imply that there was no such substance as phlogiston. Instead, he developed the theory of oxygen combustion. Priestly, however, chose to maintain his belief in phlogiston by making a different adjustment to his "web of belief." He could, for example, take the results to imply that phlogiston had a negative mass. This would account for the fact that the mass of the mercury actually increased as the quantity of phlogiston decreased.

Science provides many examples of how an experimental result can be interpreted either as a decisive blow against a theory or merely as an impetus to refine the theory in one or another way. What scientists think an experimental result shows about a theory depends on their overall view of the strengths and weaknesses of that theory. Neither of two researchers, one who abandons the theory because of the result, the other who retools it, need be unreasonable or blind to the evidence in choosing a route.

In the late nineteenth century, a series of investigations into the speed of light, which came to be known collectively as the Michelson-Morley experiment, had results with implications about the existence of the "aether." The aether was a substance that supposedly occupied all space, and its existence was thought to be required to explain how light waves could be propagated through space, since all wave transmission requires a medium. If the aether really existed, it was supposed, then Earth's movement through it should be detectable by measuring the speed of light in different directions relative to the aether. When no difference was observed, it was logically possible to abandon a belief in the existence of aether. On the other hand, a commitment to aether theory would lead one to interpret the results of the Michelson-Morley experiment in such a way as to save the theory. It was consistent with the experimental results, for example, to conclude that there was an envelope of aether, which moved along with Earth. This would explain why there was no observed "aether wind."

Quine and other holists conclude, from examples such as this, that it is possible to revise any part of a theory, make appropriate adjustments in other areas that are logically connected, and yet leave the "empirical content" of the whole unchanged. The commitments that incline scientists to go one way rather than another need not be extrascientific commitments. For example, to give up the aether theory was to be without an account of how light travels through vacuums, while merely refining the theory enabled one to maintain an elegant explanation of that phenomenon.

In the case of global models, we see that a commitment to the greenhouse effect model, for example, will allow a scientist to make adjustments that preserve this model in the face of unexpected observations. The inclination for some scientists to hold on to their greenhouse effect model, for example, in face of what some consider refuting evidence, is not an unreasonable inclination.

Why these adjustments are part of good science in some cases but not in others is explained by whether the adjustments lead to novel predictions, or whether they are just a dead end manoeuvre. Imre Lakatos divides theories into two classes: "progressive" and "degenerating" research programs. A "sophisticated falsificationist,"[4] he claims, accepts a theory as scientific, even if it needs constant saving from apparent falsifying observations, as long as it grows and makes new predictions that might lead to the discovery of novel facts. I objected above that theories like global warming can be saved by adding seemingly ad hoc hypotheses to explain away any differences between prediction and empirical measurement. However, if these addenda contribute to a "progressive" research program rather than a "degenerating" one (simply growing more cumbersome without additional predictive power), then the theory is evolving and could be considered to be a progressive research program.

THE DILEMMA BETWEEN GENERALITY AND PRECISION

As explained in chapter 2, Cartwright has argued that the sacrifice of predictive power for the sake of explanation reflects a fundamental opposition between competing virtues. One must make trade-offs between explanatory power and the descriptive adequacy needed for prediction. "Truth and explanation," she writes, "exclude" each other.[5] Idealizations collect similar things together and organize phenomena in a way extremely useful for noting general behaviours. Unfortunately, this function must necessarily be performed at the expense of details, and this results in weak predictive power.

General laws in physics, argues Cartwright, explain by using what she refers to as "*ceteris paribus* laws." These simplified laws are correct – when all else is equal. However, since all things are never equal, these laws almost never apply to an actual situation. Her argument about general laws in physics is the same as Peters's arguments about ecological models. Peters calls such constructions logical tautologies. Cartwright uses stronger language, she calls them false. She writes, for example, that the general laws of physics "do not state the truth"[6] and must be "put right" by "judicious corrections."[7]

Peters reasons much the same way. He demonstrates that there is a trade-off between broad conceptual analysis and precision. The former accommodates observations more easily, but only because it is so general that it makes no risky predictions.[8]

We seem to be stuck in a dilemma between generality and precision that scientists have been struggling with since it was first described in the Plato dialogue *Meno*. Meno's Dilemma is that it is impossible to discover what we are looking for because, in order to recognize when we have found it, we must know beforehand the very information we are seeking. We either know, in advance, what we are looking for, in which case we don't need to look for it, or else we endlessly accumulate observations without knowing whether we have found what we are looking for.

Peters's arguments above show that models can only be predictive if it can be specified *beforehand* whether they apply in any particular case. The knowledge needed to do this, however, is precisely what the model is supposed to supply. As we have seen above, global models are tautologies because ad hoc assumptions can explain away observations that do not match model predictions.

Similarly, we have seen how global models are often tuned so that certain errors are not multiplied, causing some unrealistic projections of behaviour. However, to perform this tuning we must know beforehand what behaviour is expected, and this knowledge is what the model is supposed to supply. Thus, we are in the dilemma of needing to know beforehand what to expect – in which case we do not need the model. Or else we do not know what to expect, in which case we do not know whether our model is correct. In most cases of modelling, this dilemma is escaped by a process of trial and error until the model achieves predictive success. However, in the case of global models predicting catastrophe, this escape will seldom be available.

FOUR DEFENCES OF MODELS WITH WEAK PREDICTIVE POWER

The method of creating an ideal model and coupling it with explanations after the fact is hard to give up. Models can be defended from accusations of weak predictive power by arguing that the accusations are too broad, that models can still be scientific even if they cannot be applied, that models have heuristic value, and that models are useful for historical explanation.

Accusations Are Too Broad

It might appear that Peters's argument is attacking all idealized models, but this would be a mistaken interpretation. It is the application of the model that is important. A model is only tautologous if we do not know whether it fits the world until we try it.

In defense against the tautology argument, one could point out that, despite the above criticisms, scientists use ideal models to explain things all the time. Peters's attack might seem so broad that we would have to devalue many of our most cherished laws. One might think that a parallel argument could be run against the Ideal Gas Law, for instance, yet we do not dismiss it as uninformative about the real world!

Part of what is going on here is the previously mentioned trade-off between simplicity and detail. All modellers must make simplifying assumptions to avoid getting overwhelmed with details that might be impossible to obtain and that are unnecessary for a model's purpose. Models are *intentionally* idealized representations. There is no such thing as a frictionless surface or a point source of gravity, for example. Models are intended only as imperfect approximations and "logically possible" constructions of how things might be. You cannot attack a model for being a model.

The difference, however, between idealizations like the Ideal Gas Law and those of global models is that in the former we know what adjustments are needed to apply them to real cases, whereas in the latter we do not. Cartwright writes that for an idealization to be useful "we had better know how to add back the contributions of the factors that have been left out."[9] We know how to adjust basic laws when the surfaces are not frictionless and the mass is extended in space rather than an ideal point source. However, as Peters has shown, we do not know how to apply tautological models to a real situation until after the fact, and I am claiming that global models fit this category.

Models Can Still Be Scientific Even If They Cannot Be Applied

It might be possible to defend against Peters's argument by appealing to the semantic theory of models described in chapter 2. Arguments similar to those described above have been levelled at the theory of sociobiology. It too has been accused of being unscientific because the theory's claims are routinely supported with ad hoc explanations and cannot be falsified. Paul Thompson, however, has defended sociobiology by appealing to the semantic view of models.[10] On the semantic view, a theory is a set of models that are applied to the world. Thompson concludes that the attacks strike against not the theory of sociobiology but merely its application to the behaviour of humans. He rescues the scientific status of sociobiological theory by giving away its usefulness for describing human behaviour. We might one day develop that part of the family of models that connects genetics to human behaviour, models that are needed for application of the theory to humans, but for now, he advises, we must be satisfied with a "promissory note."[11]

As promising as this manoeuvre might look, it can only provide an incomplete rescue for the case of global models of catastrophe. The application of models of global risk is the part that we are most interested in. We are not asking about the truth of the models here, but merely whether they will work. We want to know if any particular model is any good at prediction. I have no *in principle* objections against models of global systems that predict catastrophe, no reason to think that we cannot improve them some day. However, the ones we have right now are suspect, and, of course, we need to make judgments based on the ones we have now. I am willing to grant that models have scientific status, but the conclusion that they are currently unable to give reliable probabilities still stands.

Models Have Heuristic Value

Even if the predictions of global models have weak empirical teeth and the methods of indirect testing are not able to support anything more than vague predictions, the models might still be useful for other reasons. One could argue that predictive power is not the only evaluative criterion of models and that global models have heuristic value even though they are not predictive. A heuristic is a simplification that has value if it inspires further research and the production of a better model.

Modellers are not unaware of problems with the accuracy of data, and other uncertainties, so no modeller expects to prophesy the future with pinpoint accuracy. They claim, however, that models need not be exact and law-like to be of use. Rather, they need only provide a general diagnosis of long-term trends in order to be of use in guiding our actions and showing us how to avert catastrophic possibilities.

M.A. Bunge emphasizes the virtues of theories beyond their ability to generate falsifiable predictions. Bunge proposes that theories "come in various degrees of testability" and that checking the accuracy of forecasts is not the only way to assess a model's scientific worth.[12] He writes that when empirical testability is unattainable then "conceptual testability" will have to serve as our only means of judging a theory.[13] A theory is tested conceptually by examining such things as its simplicity; its internal coherence; its ability to connect various, seemingly unrelated facts; and its unity with background scientific knowledge. These criteria of acceptability are deeply lodged in the heart of scientific rationality; for example, any theory that implies a statement that contradicts a well-established thesis has a big strike against it. Models meet these criteria by representing reality in ways that can be systematically defended by reference to theories that are already accepted.

Bunge concedes that if a theory is empirically untestable *in principle*, then it would be hard to convince anyone of its connection to reality at all. The heuristic value of a model is, therefore, insufficient unless the model is "susceptible of *becoming*" empirically testable.[14] This line of thinking parallels that of Lakatos, which was described above.

Appeal to heuristic value has been made in the debate between complex and simple global climate models. The emphasis on complex models arises from the belief that the best way to represent complex systems is with complex models. It is assumed that the more complete and detailed the model, the more realistic and, hence, the better able to make reliable predictions. However, Shackley et al. have challenged this assumption about the need for complexity.[15] The use of complicated equations and coupled models might introduce larger uncertainties than necessary because the resulting model is more complex than can be justified by observations. One or two-dimensional energy balance models, radiative-convective models, and statistical-dynamical models are cheaper, easier to use, and often just as successful at predicting global averages. Complicated models attempt what might be impossible: a detailed, pseudo-realistic simulation. It might be impossible because of the nature of the environment: accuracy of data is low, and understanding of global processes is limited. Why not, then, some scientists ask, use statistical techniques specifically designed for poorly defined systems with limited observational data? Because, reply the boosters of complexity, the better prediction is bought at the expense of understanding the detailed nature of the processes. While systems models might fall short of the task now, they seem the most plausible route to success in the future; it's just a matter of working out the details. Shackley et al. suppose that this faith in the need for complexity is, however, a methodological commitment to a paradigm that many modellers do not share.

If a theory is capable of becoming empirically testable, then it can act as a heuristic – in the sense of a simplification of a complicated system that is useful as a temporary building block to further theory development. In the absence of empirical tests, the inference to the most unifying and conceptually plausible theory is the best we can do.

Models Are Useful for Historical Explanation

Global models still have explanatory value even if they are bereft of predictive power. Historical explanations after the fact can still be useful if they aid understanding. One kind of historical explanation is referred to as a "normic" explanation.[16] The normic explanation fits past events into the general framework of accepted norms. Particular events of interest are thus explained by reference to familiar rules or patterns and by showing that the observed events were not outside an expected range. This kind of explanation is nonpredictive because it cannot specify beforehand exactly which rules apply, it can only make appeal to the rules to explain events that have already occurred. Historical explanation is common with respect to explanations of human behaviour. One can explain my choice of video rental, say, by reference to my general beliefs and interests, but given my general beliefs and interests, one cannot predict beforehand which choice I will make. Similarly, scientists can explain observations, after the fact, by reference to general scientific norms. Peters illustrates this by describing how the tendency of certain birds and butterflies to glide can be explained by reference to rules of energy efficiency.

Historical explanation is fundamentally different *in kind* from scientific explanation, Peters reminds us, and must not be confused with scientific explanations in the historical sciences. Geology and evolutionary biology, for instance, use past regularities to explain observations. The difference here is that, in scientific explanations, the regularities and norms are used to show that an observation was to be expected and also why some other possible observation was not. The claim is that the observation *could have been* predicted. This is the covering-law model of explanation. Historical explanations, in contrast, cannot make this claim.

Above, I gave a detailed example of how scientists attempted to explain the discrepancy between observed global average climate temperature and the predicted average temperature. In that case, scientists were able to give plausible explanations for the difference by suggesting various interference mechanisms, but they were incapable of explaining exactly why, in this specific case, any one particular interference was more important than any other. They would not have been able to predict the observation.

CONCLUSION

Peters warns that virtues like heuristic value and conceptual coherence should not be overrated. Such attempts to find some saving virtue for a nonpredictive theory are the "defence of last resort for bankrupt theory," he claims.[17] His major concern is that ecologists often retreat, inappropriately, to the comfortable realm of abstract, conceptual theories, and models that are impregnable to criticism because the logic offers an elegant solution. Peters urges that we not make it a regular practice to relax our scientific standards by accepting historical explanation or heuristic value in place of predictive power; he reasons that appeasing our desire for predictive models with lesser substitutes would forestall the development of better, more predictive models.

More important for our concerns of assessing models that predict catastrophe, is that heuristic value, historical explanation, and scientific status are of no use for immediate problem-solving. Peters, for instance, supposes that the prevalence of historical explanation in ecology is responsible for the current crises in fisheries management around the world. Such ecological failures are a result of attachment to inaccurate models that are kept because of their irresistible logical attractiveness.[18] The fact that the models in question may undergo constant updating in response to the discrepancies found between them and the world is not necessarily a sign of progress. This may be merely "self-perpetuation and expansion."[19] He thinks that this process is potentially endless and that the only real progress that should count is progress in predictive power – and that will require knowing, in advance, whether the model is applicable to a given situation.

Interestingly, Peters himself makes an exception for the cases of global models that predict catastrophe. In these cases he accepts that historical explanation has a legitimate role and prefers a nonpredictive model over nothing at all because of the importance of the problem and the risks involved.[20] However, the great risks involved will not change the fact that historical explanation will not do the job we need done in the case of global models of the future: namely, produce predictions about the effects of our current actions. How could a historical explanation, which explains why particular events did or did not conform to the ideal model in any one case, provide strong grounds for actions intended to protect the future environment? For this, we need to know what is likely to happen, not simply why certain things did or did not happen. Unfortunately, this need is not being met by current global models because of their weak predictive power. In the absence of strong predictions, these models provide only descriptions of *possible* outcomes.

Given possibilities and not probabilities, it is difficult to decide what to do. Perhaps we should follow the humourous advice of cartoonist Robert Mankoff in one of his contributions to *The New Yorker*: "And, while there is no reason yet to panic, I think it only prudent that we make preparations to panic."[21]

PART TWO

In Policy about Global Risk, Precaution Will Maximize Expected Utility

·7·

From Models to Decisions: Science and Values in Decision-making

The problem raised by models of global systems that predict catastrophe involves two questions: what to believe and what to do. The first is epistemic: what should we believe? The first part of this book was concerned with what good reasons there are for accepting predictions supplied by models of global systems. Our beliefs about the accuracy of these predictions will remain uncertain for some time. We are uncertain, therefore, whether to take seriously recommendations for action based on model representations, even if such action might prevent outcomes we would consider catastrophic. The second question is practical: what, then, should we do? Science by itself cannot drive policy because other values must figure in decision-making. This second part of the book will deal with proposed responses to the threat of global catastrophe. What sorts of risks can we justify taking? How do we want to live? What should we do?

Suppose one accepts the thesis that models of global systems cannot provide any definite reason for thinking that human activities are contributing to some imminent global catastrophe. Does one's acceptance mean that one should support the claim that there is no good reason to act on the recommendations for avoiding catastrophe supplied by these models? Not necessarily. Despite the uncertainty of predictions generated by models of global systems, it turns out that the threats to biodiversity, resources, coastal cities, and agriculture posed by global warming, ozone depletion, population increase, or nuclear winter, though not established, have a non-negligible degree of probability.

While global models might not be able to provide definitive reasons for thinking that catastrophe is likely, they do make us aware of frightening possibilities and, moreover, supply us with guidance on how to eliminate them. Global catastrophe, however unlikely, is possible.

We do not know if we are on the brink of disaster or not. The world is large and complex, and our models of it cannot provide adequate information about the likelihood of environmental disasters. Given that we are forced to accept a seemingly intractable uncertainty, how are we to face these unknowns and make decisions? Though we may not discover a method for testing or justifying faith in our global models, since the stakes involved here force us to make a decision now about how we will respond to them, we must find a method for weighing the risks involved.

This assessment of risks requires that we make prior decisions about values and desires. Science alone is inadequate to decide for us what to do in the face of possible catastrophe. Even if models were able to provide accurate probabilities, this knowledge alone could not suggest action independently of our desires and values.

Religious beliefs, for example, will affect how one values environmental health, and, as Thomas Hurka stresses quite rightly, before a decision about how to react to the threat of global warming can be made, one must first decide what to believe about the extent of our ethical obligations.[1] Whether our sense of moral obligation extends to include future generations or inclines us to value the environment for itself rather than instrumentally or whether we are content to restrict our moral attentions to humans here and now will affect our response to perceived threats. Hurka notes, for example, that the more encompassing our circle of moral responsibilities, the more likely we would be to choose a policy of strict avoidance of global warming rather than one of adaptation to it.

Because values in addition to the epistemic value of apportioning our belief to the evidence are involved, decision-makers must take into account the general perception that the world is a fragile place, even if they judge the models scientists offer them to be low in predictive power. The visible, palpable degradation of the environment, the exponential rise in population, the loss of forests, the high rate of species extinction, the decline of fisheries, the fouling of rivers – all generate, for many people, a fear of catastrophic environmental collapse. Many authors who are critical of models of global risk nonetheless conclude that we should be concerned with the state of the natural environment because of the history of environmental crises and the apparent trend toward increasing environmental stress. We know from measurements of ozone depletion that human action is capable of causing global changes quickly, and it is possible that the environment is not as resilient as many people might once have assumed it to be. We have seen that it is possible that some activities are putting Earth at risk of some catastrophe. Given the scientific uncertainty and yet the need to decide what to do about the risk of global catastrophe, there is a need for some strategy of decision-making under circumstances of scientific uncertainty and great risk.

The issue here is whether caution is, all things considered, a good strategy. Is precaution backed by either moral reasons or prudential reasons (or both)? Are there stronger moral or prudential reasons against precaution? The decision whether caution is a good strategy will involve problems other than just ones concerning how to measure the reliability of predictions generated by global models. As well as trying our best to assess the dangers and benefits of our projects and plans, we must make decisions about whether it is morally permissible to gamble with such high stakes. Should we make minimizing risk a priority and reduce the risk of catastrophe as much as possible no matter what the cost? Or, should we attempt to weigh the benefits of additional safety against the costs and try to maximize expected utility? It will become clear in this discussion that decisions about what to do in response to threats of global catastrophe could not be resolved even if we were to solve the scientific questions about uncertainty and testability of models of global risk. The immensity of the risk, the distribution of the risk, and the desirability of various trade-offs will determine our attitude toward any particular risk.

We do not actually have attitudes to risk per se. In any risky action, one must decide which trade-offs one is willing to make. That an activity – say, hurtling down the road at high speed – is risky is one feature of it that influences our attitude toward it, but not the only one: it will enable us to reduce the boredom of being on the road between Ottawa and Toronto, it will get us to where we want to be faster, it is exhilarating, and so on. Hurtling down the road might be likely to injure us, but, hurtling down the road also has much in its favour. Similarly, there is a risk of catastrophe from global warming, but also much to say in favour of the industrial, agricultural, and personal activities that might result in global warming. The possibility of catastrophe does give us reasons to act cautiously. However, these reasons might be overridden by others. An examination of such trade-offs might reveal that a desire to preserve the rain forests to maintain a carbon sink is in conflict with the desire to ease the poverty of developing nations and their expanding populations. However, just what is the minimum we must do to avert catastrophe – just how many of our present ways and present values (e.g., development, fun, jobs, comfort, mobility, freedom, democracy) we can keep – is deeply controversial.

DECISION THEORY[2]

Obviously, the difficulty of decision-making under uncertainty is not unique to global modelling. Decisions about economics, for example, must also be made under considerable uncertainty, and a huge literature on the rationality of decision-making exists.

Decision theory requires that we conceive of decisions in terms of matrices. A decision matrix is formed by creating an exhaustive set of mutually exclusive actions from which we must choose and by listing them against another exhaustive set of mutually exclusive possible states of the world. We then make a value assignment for each possible outcome (each action/state pair), either by ranking or by an arbitrary unit of comparison. Our decision about what to do is then determined by comparison of these value assignments. (See the example on page 83, which uses an arbitrary unit of comparison ranging from 0 to 100, with 100 being the highest value.)

In situations of certainty, where we know which possible state of the world obtains, we simply choose the option with the highest value assignment for that state. In the decision matrix on page 83, if we know that the summit of Mount Everest is stormy we would prefer not to climb; if the summit is experiencing fine weather we prefer to climb.

More likely is the situation where we do not know which state of the world obtains. Nonetheless, many decision problems have one best course of action no matter what happens – a so-called "dominant" alternative. Someone might prefer to climb no matter what the weather. However, it is often not the case that there is a unique optimal decision. Sometimes the preferred action varies with the state – as in the above example.

In the absence of a dominant alternative, the favoured approach of Bayesian decision theorists under normal (noncatastrophic) circumstances is to calculate the expected utility of each course of action we believe open to us and then choose the course of action we believe has the highest expected utility. Expected utilities are calculated by multiplying

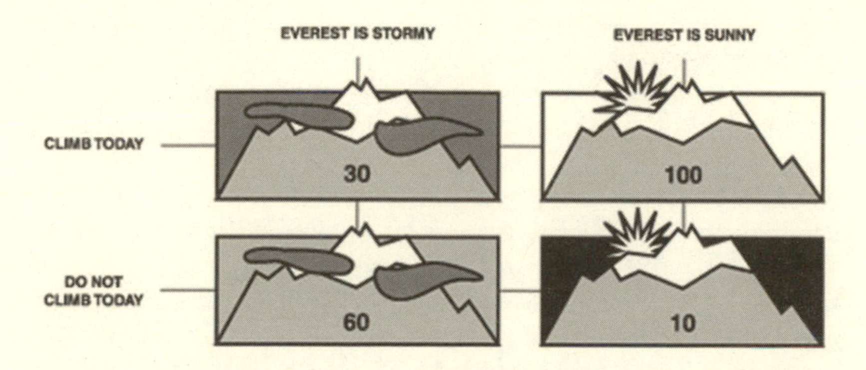

the value assignments of the matrix by the estimated probability of each state's coming to pass, and then adding these products for each possible choice. If the probability of stormy weather on Mount Everest is 50/50, then the expected utility for climbing today would be (30 x 0.5) + (100 x 0.5) = 65, while the expected utility of not climbing would be (60 x 0.5) + (10 x 0.5) = 35. So, given this assignment of values and probabilities, the alpinist should prefer to climb today.

This decision procedure under known probabilities is referred to as decision under risk, and is the foundation for cost-benefit analysis. Such a calculative procedure ensures the most efficient decision-making because the benefits and losses are weighted according to their likelihood.

Sometimes, unfortunately, the probabilities are uncertain or unknown. The accepted strategy in the face of uncertainty is to avoid making any major decisions until more information has been gathered and the uncertainty has been reduced. Nicholas Rescher comments that cases of uncertainty "call for further inquiry into the processes at issue ... Deliberation is premature ... [and] further investigation is needed."[3] Ronald Giere also recommends suspended judgment: "the rational attitude in these cases [of uncertainty] can only be one of suspended judgment."[4] This is a common reaction to the underdetermination by evidence of models that generate predictions of global risk, and we hear Giere's conclusions echoing in the newspapers. Richard Worzel of *The Globe and Mail* writes, for example, "Should we ... be spending billions of dollars to restrict the emission of greenhouse gases? I believe we are too ignorant to risk so much of our financial resources on so uncertain an outcome."[5]

However, we cannot enjoy that luxury in the cases of global risk. If we accepted this advice in cases of global risk involving potentially irreversible catastrophes, the very purpose of the decision would be defeated.

Simply waiting is unacceptable because, given the often catastrophic nature of the predicted consequences, we are forced to make a decision before waiting for information – otherwise it may be too late. The nature of some of these suspected catastrophes is such that, if the models are right, preventative action is needed immediately, and any delays might make the catastrophes inevitable. If the predictions are correct, then any delays in action will only make the situation more difficult (if not impossible) to correct. It is suspected that many of Earth's ecosystems–coral reefs, mangrove, forests, grasslands–may be nearing a threshold point of irreversible damage, and as Jon Elster reminds us, "Destruction of the environment is often nearly impossible to undo. 2000 years ago the land around the Mediterranean (including much of the Sahara) was green and fertile. Excessive cultivation and deforestation rendered it barren."[6]

GIERE'S NATURALISTIC ACCOUNT OF DECISION-MAKING

In *Explaining Science*, Giere offers a descriptive account of how scientists actually make judgments about which model best represents the world. This is a naturalistic account, he argues, because it is based on investigation into natural cognitive patterns of judgment shared by all humans, rather than on an idealized rationality. He argues that in science as actually practiced, scientists come to accept or reject theories and models on the basis of a satisficing decision strategy that involves "both individual judgment and social interaction."[7] Giere's account concerns decisions about which model to accept, not about what policies to follow, but I think his account can also be successfully applied to questions about what to do in response to risks of global catastrophe.

Giere rejects the Bayesian decision theory account described above because he rejects the claim that humans are natural Bayesian calculators. He presents the results of several well-known psychological experiments (Kahneman, Slovik, and Tversky) that show that even those with specialized training in statistics and probability can make common errors in judgment, because, like all of us, their natural cognitive abilities are designed in such a way that they can be easily misled in unusual circumstances. These experiments show, for example, that our natural judgment can be directed by notions such as representativeness, even when this concept is inappropriate to the problem. The mistake here is that we judge certain problems to belong to a class that is representative of some general pattern and then make estimates of probability based on perceived similarities between the problem and the representative class rather than the frequencies given in the problem. Thus, we get people mistakenly asserting that it is more likely that some person is both a bank teller and a feminist, than a bank teller alone, because they have been given a description of that person that they believe is representative of the class of feminist bank tellers. Other examples illustrate other cognitive biases, such as the tendency to ignore baseline information.

How, then, do scientists manage to succeed? Rather than attempt to provide criteria for making rational choices about theory acceptance, Giere responds that scientific decision-making can best be explained in terms of a satisficing decision theory, rather than as the result of logical inference. Herbert Simon invented the strategy of satisficing to deal with decisions in situations of bounded rationality, where probabilities are uncertain. We can imagine that scientists create a decision matrix like those described earlier. The evaluation of how well some model fits the real world is based on the epistemic value attached to keeping one's beliefs in line with the evidence, but also on the value of other interests – such as a political commitment to a certain theory, or simply familiarity and expertise with it. The beauty of this account, says Giere, is that it deals with the problem of the place of values in science. Extra-scientific interests are included right from the start because they are necessary to obtain a value ranking for the matrix construction. In Giere's words: "A decision matrix literally requires that there be a value rating attached to the outcomes of the decision matrix. There is thus no question whether values enter into the choice of scientific models, or even how ... the only question is what values."[8] It is a satisficing account because it does not appeal to some absolute standard of rationality for making decisions. One need not believe the chosen model, but merely accept it for now or take it as the best working hypothesis.

Despite the intrinsic presence of values in theory choice, scientists often agree, Giere reports, because sometimes the evidence in favour of one model is so clear that epistemic values "overwhelm" all others.[9] He gives a lengthy description of the rise of plate-tectonic theory and its eventual acceptance by most geologists in order to show that when data are uncertain, political commitments can decide the status of any particular theory. In this case, the professional interests of geologists in the 1920s led to a rejection of the theory of continental drift. Giere comments that this rejection was rationally neither warranted nor unwarranted. There are no principles of rational warrant, he claims, but only good and bad strategies of decision-making. As evidence mounted in favour of the plate-tectonic theory, the epistemic interest in being correct outweighed any other interests scientists might have had. The probability that plate-tectonic theory was correct – given the magnetic evidence of ocean floor spreading, geographic resemblance of coastlines, and other evidence – became large enough that it could not be ignored in favour of other interests that scientists might have had. This development parallels the acceptance of the theory of ozone-depletion due to the action of CFCs in the stratosphere.

Giere's decision-theoretic account of science has to do with choosing what model or theory to accept. For my purposes, I want to map it onto decisions about what to do. When trying to decide among the strategies of action presented in part I, we might be able to choose satisfactory outcomes even though we cannot calculate expected probabilities. Just as a criminal trial serves to further goals in addition to finding the correct answer to the question of guilt or innocence (e.g., it serves to show that justice is being done, it provides a deterrent), our prior political commitments will affect our choice of response to models of global risk because we are trying to further other goals besides the goal of getting the science right. A choice of strategies made according to this satisficing strategy rather than an attempt to maximize expected utility can provide room for ethical and political interests.

·8·

Decision-making Under Uncertainty and High Risk

When probabilities are very uncertain or entirely unknown, the preferred strategy among decision theorists is to avoid deciding what to do until more information has been gathered and the uncertainty has been reduced. However, as I said in the previous chapter, were we to adopt this strategy in cases of global risk involving potentially irreversible catastrophes, the very purpose of making a decision would be defeated. In such cases we are forced to make a decision one way or the other because we cannot afford not to decide, and we cannot afford to wait for more evidence because, as we saw in part 1, such evidence might be slow in coming – if it ever does come.

We must face squarely the problem of making momentous decisions under uncertainty. If global models are correct in predicting that disaster will occur unless we take preventative action, we would prefer to take preventative measures. However, if predictions are erroneous, then we would prefer not to go to the trouble of changing our behaviour and living in ways we like less than our present ways. We have one central choice here. We can take a precautionary stance toward the possibility of global catastrophe, and do what we can to eliminate that possibility, including abandoning a lifestyle we find comfortable. Or we can take a merely careful stance, and accept some risk of catastrophe in pursuing comfortable lives. The dilemma we must confront in this choice is that between overcautiousness on the one hand and recklessness on the other.

This chapter will outline the dilemma that the possibility of global disaster creates, and then discuss two possible strategies. Epistemic and practical conservatism is the strategy of not reacting to suggestions of catastrophe unless there is sufficient proof that one has an unacceptable probability of occurring. Unfortunately, this conservatism in environmental policy-making might lead to catastrophe. The precautionary principle, on the other hand, urges preventative action without the need for scientific proof. However, this policy may be too costly and too difficult to implement. Once this dilemma has been outlined I will discuss, in the next two chapters, possible ways of resolving this dilemma.

The nature of this dilemma is captured in the defining moment of Joseph Conrad's *Lord Jim*. Jim was an officer on board the *Patna*, a ship carrying eight hundred "pilgrims of an exacting faith" to Mecca. The *Patna* struck something in the water and, thinking that the ship was about to sink at any moment, the entire crew abandoned ship quickly. The ship, however, did not sink but was found adrift days later.

In the face of uncertainty about the state of his vessel, Jim abandoned ship when the ship was not actually sinking, and thereby brought dishonour to his whole life. By jumping ship, Jim committed one of two possible errors – he acted unnecessarily. The complementary error, of course, is not acting when it is required – to stay on board when the ship really is sinking. Almost everyone would agree that, from Jim's point of view, the error Jim actually made was preferable to the error that might have been – in that case he would have drowned. (The remaining possible outcomes would be for Jim to get things right – to jump ship when ships are sinking and to stay on board when they are not.)

This decision matrix[1] for some global catastrophe is parallel to Jim's case:

The possible errors here are logically similar to each other – to take action to prevent, say, asteroid strikes when none is actually required; or to take no protective measures and get blasted by a speeding chunk of outer space flotsam that might easily have been deflected with a well-placed, though no doubt expensive, missile strike. (Of course, the launching of such missiles might present an even greater risk – but this is just an example.) One estimate is that the strike of an asteroid with catastrophic potential – roughly one kilometer in diameter – will occur once or twice every million years.[2] Just as in the case of Lord Jim, most people would agree that the former error (taking needless action) is preferable to the latter (taking no action when action is required.)

However, this is not exactly the choice that we are forced to make. There is a chance that we might get everything right and make neither

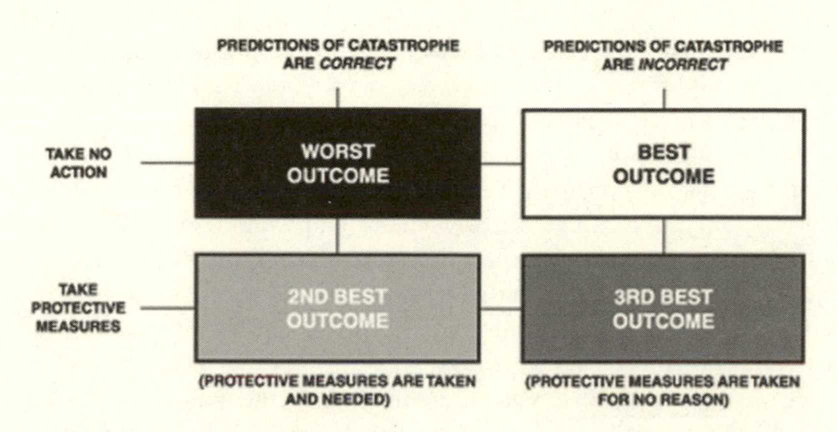

error. Remember both that we do not know at this time whether our ship is, in fact, sinking and that even the more preferable of the two errors is not without its cost. When Lord Jim ruminated on the events that took place that night, all he could think of was that, by jumping unnecessarily, he lost an opportunity for advancing his reputation as a hero in the face of adversity. Besides opportunity costs, there is a more direct cost. If we jump ship every time we suspect trouble, we will spend a lot of time in the water. There is a multitude of catastrophic possibilities awaiting anticipatory action. In *The End of the World*, John Leslie gathers together a frightening catalogue of potential catastrophes and details numerous now-familiar risks, including warfare, diseases, and overpopulation, as well as unfamiliar or often overlooked risks ranging from asteroid strikes to nanotechnology run amok and universe-annihilation resulting from misadventures in the physics lab. If we take action to avoid every suggested catastrophe, the costs will be substantial.

TYPE-I AND TYPE-II ERRORS[3]

This dilemma of deciding between a risky opportunity and a more cautious course of action has been described in the literature in terms of committing "type-I" or "type-II" errors. A type-I error occurs when we reject a true null hypothesis. A null hypothesis is a claim of no effect. For example: 1) my ship is *not* sinking, or 2) curtailing the use of fossil fuels will have *no effect* on global warming.

Suppose the above hypotheses are true and yet, in our uncertainty, we reject them (that is, we think the ship *is* sinking, and believe that taking protective measures *will* have an effect on global warming). We would, then, take measures that are unnecessary. The error occurred in overemphasizing the risk of catastrophe, for this caused us to lose opportunity and resources for development while receiving no benefits. A type-I error is an error of *exclusion of truth*. By failing to believe something that is true, we miss an opportunity.

A type-II error, on the other hand, occurs when we fail to reject a false null hypothesis. Let us use the same null hypotheses as above except this time assume that they are false (that is, the ship actually *is* sinking, and the taking of measures *will* have an effect). Now suppose we make the mistake of failing to reject these null hypotheses (that is, we accept them as true when they are really false), then we will not take protective measures when, actually, taking them would have paid off. A type-II error is an error of *inclusion of falsehood*. We included in our belief set that which was false and hence underestimated the risk of catastrophe. In this case, rather than losing opportunity, we lose what we already have.

This jargon of type-I and type-II errors can get confusing because, with a little rewording of the null hypotheses, one can change the "propositional content" on a type-I error into a type-II error, and vice versa. Thus, I believe it reflects a mere grammatical rather than a logical difference. For example, we could word the null hypothesis in one of the following two ways: 1) curtailing the use of fossil fuels will have no effect on global warming, or 2) continuing use of fossil fuels will have no effect on global warming. The wording of the null hypothesis can thus reverse the type-I/type-II classification. Despite the widespread use of these terms in the literature, I will forgo the use of them and talk instead about conservatism versus nonconservatism.

EPISTEMIC CONSERVATISM

The dilemma described above is about errors of action. However, the same logical problem exists with respect to beliefs. In order to avoid jumping on to every theoretical bandwagon that comes along, the institutions of science have adopted the strategy of treating new theories, models, and hypotheses as suspect until proven otherwise. In order to maximize true belief and minimize false belief, it has become standard practice in science to take steps that ensure a low probability of mistaken acceptance of a theory or hypothesis. Scientists want to avoid incidents like the Piltdown Man hoax.[4] The Piltdown Man hoax was a very famous case of scientists mistakenly believing something without sufficient proof. From 1912 to 1953, many scientists accepted the Piltdown skull as evidence of a "missing link" in the evolution of humans from apes. When it was discovered that the skull actually consisted of an orangutan jaw and a modern human cranium, fifty years of scientific research had been infected with false assumptions arising from the false data. To minimize the chances of such mistaken acceptance, the burden of proof is on those who wish to make assertions and add to the body of scientific belief.

This is also the motivation that lies behind Ronald Giere's "improbability condition" in his criteria for a good test mentioned in chapter 3. In order for a theory or hypothesis to be accepted, it must make a verifiable prediction that is *highly improbable* given everything else known at the time.[5] This ensures that not just any plausible hypothesis will be brought into the fold of accepted belief. Nothing that could be explained equally well by other means will count as evidence for a theory. This high standard of evidence makes it difficult to make the error of including falsehoods in a belief system. Conversely, it makes the alternative error more common – namely, the nonacceptance of a truth.

When one or the other type of error is unavoidable, the institutions of science follow the maxim of epistemic conservatism: "Hypothesis-testing in science operates on the basis of limiting false positives (assertions of effects where none exists)."[6] The practice of science demands a high degree of confirmation and replication of an experiment before its acceptance, and scientists are taught to be unwavering skeptics. Scientists prefer to wait for certainty before adjusting their belief system because science, as an institution, would degenerate if it were not epistemically conservative.

J. Cargile notes that it is a "common principle" that the burden of proof always lies with those who make assertions; for example, the assertion that it is possible to square a circle requires proof.[7] If we adopt this pattern of reasoning for our dilemma about what to do when faced with a risk of global disaster, the burden of proof will lie with those who assert that there is such a risk, rather than with those who want to continue business-as-usual. Thus, to put faith in warnings that the sky is falling – warnings that cannot pass Giere's rigorous testing criteria – runs counter to the logic of normal scientific practice. Epistemic conservatism, the asymmetrical preference for errors of truth exclusion over errors of falsehood inclusion, when transferred to our problem of risk of disaster, will protect us from overreacting to every possible doomsday theory. However, we will make ourselves vulnerable to the other error. That is, while we decrease our chances of reacting to false alarms and avoiding, unnecessarily, opportunities for development, this comes at the cost of increasing our chances of proceeding with practices that actually will lead to disaster. Demands for further proof will reinforce institutional inertia and inaction. The environmental policy of the United States illustrates this reluctance to accept warnings of catastrophe; for example, the US government did not sign the proposal to limit greenhouse gas emissions proposed at the June 1992 Rio summit.[8] In the words of one policy official "Scientific uncertainties must be reduced before we commit the nation's economic future to drastic and potentially misplaced policy responses."[9] In the case of doomsday forecasts, however, reserving judgment and doing nothing could be more costly. The longer we wait to address environmental crises, the more difficult it becomes to correct them.

It might be objected that scientific reasoning, which attempts to describe the world, should not be thought analogous to practical reasoning. When Lord Jim jumps ship, he *does* something, whereas scientists merely reserve judgment. Reasoning about what to do has many features different from reasoning about what is the case.

However, intellectual caution transfers to practical caution as well. Parallel reasoning can be found in the legal system, which does have to do with practical reasoning. The legal system is designed to resist the assignment of guilt where none exists. We minimize the chance of convicting the innocent, even though it is acknowledged that this will result in an increased probability of nonconviction of the guilty. (It is interesting to note that in the French legal tradition the burden of proof is reversed, and one author, A.R.D. Stebbing, speculates that this difference in legal reasoning might be reflected in the environmental policy choices of that country.[10] In at least one case, involving marine pollution, the decision to take precautionary action was made, he claims, on the basis of much less evidence than that required elsewhere. Thus, practical reasoning with respect to the environment parallels the practical reasoning of legal courts for both France and the United States even though this reasoning is different in each case.) What we need to know is whether there are any good prudential or moral reasons for escaping epistemic conservatism when confronted with the possibility of global disaster.

COST-BENEFIT ANALYSIS

"A man sits as many risks as he runs," Thoreau asserts.[11] If the risks are uncertain, and yet we cannot afford to suspend judgment, why not just do the best we can with the strategy of calculating expected utilities and trying to maximize them? In fact, taking calculated risks is often seen as an essential and laudable aspect of human nature. Risk-taking can pay off and bring rewards that are unavailable to those who are cautious.

Cost-benefit analysis requires estimating the value of various outcomes in a decision matrix, as described in chapter 7. For example, the outcome of catastrophic global warming will have a lesser value than the state of benign global warming. The costs of abating greenhouse gas emissions will also affect the final value of the outcome. In the case of a decision about marine pollution, the costs of constant and extensive monitoring of the marine environment for harmful effects of toxic dumping, together with damage control costs, might outweigh the benefits of dumping.

In cost-benefit analyses of noncatastrophic risk, the risk of an undesirable outcome on some occasions might be accepted because it is outweighed by the chance for benefits. We accept these risks in some cases because, when we bet on probabilities, we usually come out ahead in the long run. In most cases of decisions under known probability and noncatastrophic risk, one can calculate the odds and the cost-benefit ratio and bet on the probability that, in the long run, things will work out. If we have made a mistake in judgment, we can learn through trial and error, and thereby improve future outcomes. Also, most risks can be compensated for or insured against, so that a bad outcome is only a setback and not a total disaster. The overall strategy is one of maximizing average expected utility. However, as will be emphasized below, there are features particular to catastrophic risks that make this long-term averaging inappropriate.

The method of cost-benefit analysis is inadequate to the task of deciding what to do in the situations of extreme risk described in the zero-infinity dilemma. The uncertainty of the risks in question is high, and the consequences of failing to change our behaviour could be catastrophic. "Catastrophe dances an intricate minuet with improbability, and the virtual infinitude of the one plays off against the effective zerohood of the other," writes Nicholas Rescher.[12]

The first problem has to do with the "infinity" side of the dilemma. The procedure of measuring the weight of an outcome by the product of its probability and assigned value cannot differentiate between two very different situations. For example, it treats equivalently the small chance of great catastrophe and a large chance of a small loss, whereas in decisions about global catastrophe we should like to differentiate between these two situations. When the probability of an outcome is very small, but the possible negative consequences are so enormously disastrous and potentially irreversible, then even the slightest risk of their realization raises concern, and the idea that benefits will be maximized in the long run becomes empty. If no recovery is possible there is no long run. These are one-shot deals, not casino games where we can afford to take some losses here and there in the hopes of coming out ahead at the end of the day. To take such a risk can be likened to playing a game of Russian roulette.

Thus, the risks we are considering are in a special category because of the high stakes and the irreversibility of the outcomes. If some action is potentially the cause of irreversible damage, there is no chance for improved decision-making over time and no room for experimenting until we see how it goes. It might be the case that the only way to determine the limits of environmental capacity is to go beyond them and thus destroy that capacity in the process – a kind of environmental version of the Heisenberg uncertainty principle (where in order to find out something about subatomic particles you have to interfere with them, and thus destroy the information you are seeking). Because of long delays in environmental feedback we may not be able to make adjustments and corrections quickly enough to reverse any damage done. The longer unforeseen consequences remain unchecked, the more difficult it is to make corrections. Because of time lags in many systems, we must anticipate results and make decisions well in advance of the actual occurrence of the projected outcomes; for example, in the case of population growth, Meadows et al. claim that the crisis is imminent – the exponential nature of growth in total world population means that we might already be in a crisis. The next doubling of the population will be even harder to accommodate than the last. Similarly, it is estimated that even if we were to stop producing CFCs today, the concentration already present will continue to deplete the ozone layer for another forty years. The longer the delay in taking action, the fewer the options that are available to us and the lower the chance of achieving the most desirable outcome.

Another reason why expected-utility analysis is inapplicable to cases of global risk is connected to the "zero" side of the dilemma. Uncertainty is omnipresent. Rescher has described three possible types of uncertainty[13]: uncertainty of probability that some particular outcome may result (undetermined or underdetermined), uncertainty in evaluation, and uncertainty about the nature of the outcome itself. In the absence of sufficient information about how either to determine the probabilities or to make quantitative assessments of the values involved, we cannot expect to perform some algorithmic kind of cost-benefit analysis.

Even if we decide to do the best we can with the cost-benefit approach, often our knowledge of the probability side of the equation is very limited. There is simply no way to get meaningful estimates of probability for the occurrence of some events involving global risk – we have insufficient experience with these problems, and hence the probability of their unfolding remains uncertain. We might hope to rely on expert judgements for meaningful estimates of risk. However, writers such as Shrader-Frechette note that many past expert attempts to provide probability judgments have been simply inaccurate.[14]

One might just assume equal probabilities for each possible outcome. However, this equiprobability assumption can lead to serious errors.[15] To illustrate this problem, it is helpful to recall Pascal's Wager for the belief in God, and the standard objection to that argument. Pascal noted that we have two options – we can believe in God, or not. He also assumed that there are two possible states of the universe – either God exists, or not. However, this is not exactly correct. Pascal could have chosen several different possible states of the universe depending on which type of god is postulated to exist. In order for cost-benefit analysis to supply a unique decision, it is necessary to list an exhaustive set of mutually exclusive possible states. Pascal limited the number of states involved in his choice to two, whereas he could have included the possibility of a multitude of gods, or compassionate gods, or unforgiving, hellfire and brimstone gods. An expected-utility calculation based on the assumption of equal probability for each of only two states will result in different advice than a situation of equal probability for each of four or ten states. Assuming an equal probability for only two possible outcomes – either a catastrophic future or a noncatastrophic future – might make the same mistake. We might be able to devise a dozen possible future states with varying degrees of catastrophe. The assumed probability of the worst-case catastrophe for this scenario would then be lower than in the two state case. This changes the value of the outcome and might change the decision.

Faith in our capacity to weigh risks against benefits might also be unwarranted. The "balancing risks" argument assumes that we are routinely successful at taking calculated risks. Unfortunately, the history of environmental disasters shows that we have not been so successful. We do not, in fact, always weigh the harms properly; we often underestimate them.

Cost-benefit analysis also requires that we find some common measure of comparison so that it is possible to weigh harms against benefits. Our individual psychological attitudes and expressions of value make comparisons of risk in social situations extremely problematic. In order to compare and rank outcomes, you have to have some common standard against which to measure them. The most common approach is to rank things by how much we would be willing to pay for them. However, this method cannot capture the way we psychologically express our judgments about value. People might claim that they think preserving biodiversity is more important than building highways, but they are not willing to *pay* as much to preserve biodiversity as they are to build highways. The way we express our values is not expressed by a calculative method that attempts to provide the optimally efficient solution. The method of cost-benefit analysis assumes that we could calmly estimate risks and benefits as if this analysis were some neutral procedure that could achieve a balance between the two. However, Mark Sagoff argues that this completely overlooks the kind of value decisions that must be made before we even start to play this weighing game.[16] For example, to accept the results of risk analysis, we have to assume that we are willing to accept the most efficient, optimal solution, but often we choose not to balance risks in the most cost-effective manner possible. Sometimes, Sagoff explains, we decide to reduce the risk to the minimum achievable, no matter what the cost; or we spend millions on expensive rescue missions with little chance of success, when the same money could save more lives if spent elsewhere. The point is that the final outcome is not all that matters to us, and therefore we might decide to eliminate risk on principles other than those that can be expressed in terms of cost efficiency.

Even if we could assign relative values to our moral ideals, or to human rights, or to the intrinsic worth of things like the preservation of biodiversity and nonpolluted environments, it is often impossible or undesirable to compare values and rank them in the way required for cost-benefit analysis. The trade-offs between longevity and quality of life, between the total magnitude of a risk and its distribution among individuals, and between the rights of individuals not to have risks or duties imposed on them and the potential communal disaster of not imposing these risks or duties cannot simply be decided according to some scale of long-term efficiency.

Advocates of cost-benefit analysis argue that regardless of how difficult such ranking is, it is nonetheless a daily practice. The limitations and difficulties are clear to everyone, but there seems to be no reasonable alternative. At least efficiency provides a check against the whimsy of the decision-maker and can be pursued systematically. Cost-benefit analysis might at least provide *some* guidance for action and can help us perform a necessary, albeit unpleasant task. When faced with uncertainty, some decision theorists simply resort to subjective judgments, and experts are asked to give a considered opinion about the extent of the risk involved.

Ian Hacking, however, has rejected the procedure of assigning subjective probabilities as "an imitation of reasoning" that can lead to a false sense of knowledge.[17] If we are to be intellectually honest, we cannot assign probabilities if we have no experience with the risk in question. For the cases of global risk at issue, the probabilities are not remote – they are unknown. Decision-making in cases of low-probability is not the same as in cases of uncertainty. To illustrate his point, Hacking gives an example about the early use of X-rays and the ignorance of the danger involved. To assign a numerical probability to this unknown would have "made no sense."[18] It is not as if the probability was low, but rather that it was incalculable. "There are no probabilities to enter into our calculations," he writes.[19] Utilitarian calculations require inputs that are not present in cases of uncertainty.

Using similar reasoning, J. Hill has questioned the validity of risk assessments of the release of genetically modified organisms into the environment.[20] Specifically, she suggests that the hazard might be too novel to imagine, and that therefore subjective probabilities arrived at by analogy with cases of introducing exotic species are unjustified.

TWO STRATEGIES: RISK-TAKING AND PRECAUTION

We find ourselves in a situation of uncertainty: we cannot assign probabilities to the chance of global catastrophe, and we cannot foresee the consequences of our environmental policies. We can neither tell whether our actions will succeed and result in the consequences we intend, nor foretell what unintended consequences there might be. (For example, by discouraging development of nuclear power stations, we might have indirectly promoted global warming.) We want neither to burden our economic engines with the addition of unnecessary safety precautions, nor to trap ourselves in an environmental disaster from which there is no escape. There are two basic strategies for coping with uncertainty, each displaying different commitments to caution and risk-taking. Policy-makers can dismiss fears of the unknown (or at least not give them much weight) and take risks, or they can act with precaution.

Risk-taking

In the dialogue *Apology*, Plato relates Socrates's attitude toward death. One should not fear the unknown, Socrates argues, because we know not whether it be good or bad. However, Socrates adds, we know for certain that it is wrong to circumvent the laws to which one has been committed one's whole life (and thus we know it is wrong for Socrates to flee his death sentence):

> To be afraid of death is only another form of thinking that one is wise when one is not; it is to think that one knows what one does not know. No one knows with regard to death whether it is not really the greatest blessing that can happen to a man, but people dread it as though they were certain that it is the greatest evil ... But I do know that to do wrong and to disobey my superior, whether God or man, is wicked and dishonorable, and so I shall never feel more fear or aversion for something which, for all I know, may really be a blessing, than for those evils which I know to be evils.[21]

Plato represents Socrates as questioning the soundness of acting to avoid a feared but uncertain outcome – death – at the cost of something tangible and immediate – Socrates's respect for the law. As with any analogy, there are also points of disanalogy. I do not want to portray Socrates as a utilitarian. However, I want to draw attention to the analogy between Socrates's dismissal of the unknown and the dilemma of uncertainty. A modern decision-maker might argue that, when we too are forced to make a decision about something momentous (for example, how to respond to the possibility of global catastrophe), it is foolish to make large efforts to avoid something we fear, if that something remains a vague and uncertain threat, at the expense of unpleasant and costly consequences of which we are certain. Just as Socrates does not know whether death is good or bad, we do not know for certain whether global warming will be catastrophic. It might not be advisable to invest scarce resources in avoiding hypothetical, vague threats about the future because these resources might be better spent on more concrete problems we are experiencing here and now.

This is the strategy of practical conservatism. Do not react to perceived threats, is the advice, if they have not met rigorous standards of rational acceptability and if reacting to them comes at a great cost.

This strategy might be connected with optimism. A committed optimist would argue that we should have faith that scientists will be able to fix problems if and when they occur. It might be possible to adapt to large changes in the environment. Many people are already used to avoiding the sun, not exercising outdoors on days of particularly high pollution, and drinking bottled water. In Thailand parents routinely keep their babies above ground level because of high concentrations of lead. While this environmental degradation is unpleasant, one can adapt to it and it is possible to weigh this negative value against the benefits of industrial production.

There are also plenty of proposals for adapting to larger problems. We can build sea-walls in response to rising sea levels; we could cultivate massive quantities of plankton in the oceans to absorb carbon dioxide; we could build satellite lasers to break down CFCs in the stratosphere; we could devise "satellite-clouds" that cast enormous shadows on Earth's surface; we could fire a continuous stream of frozen ozone bullets into the stratosphere; we could float an immense blanket of white styrofoam chips on the ocean's surface to reflect sunlight, thereby cooling Earth's temperature.[22]

The belief that these mad-scientist schemes might actually work displays a hubris not often seen outside of scientific utopias, and it completely ignores problems of unequal distribution of abilities to pay for such adaptations, as well as the inequity between those who reap the benefits and those that face the risk. Nonetheless, some people argue that some risks are worth taking. We cannot argue simply from ignorance of the potential consequences to a cautious position of limiting our actions, because ignorance is neutral on this question – it can provide neither a favourable nor an unfavourable recommendation.

Precaution

The nineteenth governor general of Canada, Georges Vanier, was prone to give money to almost anyone who asked him for some. When his children suggested that some of these beggars did not really need the money and that, in fact, he was a chump, he replied that he would rather make the mistake of giving money to someone who did not actually need it, than make the mistake of not giving money to someone who actually did need it.[23]

When faced with uncertainty, the strategy of the *Apology*, above, urges that we should not act out of fear. The strategy of precaution, however, advises protection against the worst possible outcome – just in case. Even if the probability of catastrophe is known to be extremely low, this safe route is chosen because some risks are never worth taking. The alternative to precaution is to wait to see whether something goes wrong before reacting. However, it may then be too late to set things right. Why wait to see whether you will fall prey to lung cancer before giving up the smoking habit? It is better not to smoke in the first place. We know from experience that precautionary action with respect to environmental problems is most effective at early stages of development when, unfortunately, the evidence of any environmental damage is usually most difficult to find. Because of the enormous scale of some human activities, the method of monitoring the environment, coupled with after-the-fact correction attempts, might no longer provide sufficient environmental protection. The world is being degraded and destroyed because we wait for evidence of harm before we take action and implement control measures. We know that the degradation is the result of human activity. The problem is that we do not always know exactly which activities will be hazardous, so we usually adopt the stance of epistemic and practical conservatism. Thus, little action is taken to address potential environmental threats until after the fact when there is definite evidence of harm. According to N. Bankes, "customary international law" is not able to cope with threats of global disaster because it was designed for localized problems, and requires the establishment of a definite causal connection between some human action and actual harm done before ordering any remedial action.[24] For the most part, the "assimilative" or "adaptive" capacity principle has been accepted. This assumes that the environment is capable of absorbing some pollutants and that we need not take protective action for the environment until a monitoring program reveals that this capacity has been breached. Because of the high demand for certainty, even strong suspicions of environmental harm are deemed inadequate to justify legal restrictions. An overly permissive attitude to development allows some industrialists to discharge effluent into the sea without even knowing exactly which chemicals they are dumping.[25]

The history of environmental disasters, however, has prompted in many an attitude of caution and a desire for pre-emptive action. Our experiences with chemicals such as mercury, PCBs, DDT, and CFCs have shown that we are capable of causing large, persistent, expensive and harmful damage to the environment. The disposition to prefer to err on the side of caution when the stakes are high is becoming more common in international agreements concerning possible threats to the global environment. It has come to be known as the *precautionary principle*. The precautionary principle attempts to provide a philosophical ground for pre-emptive action in cases of uncertainty. Adopting a cautious attitude reverses the epistemically conservative stance and instead takes warnings of catastrophe seriously until proven otherwise. Since irreversible catastrophes cannot be compensated for after the fact, it is argued that they should be completely avoided. In *Risk*, Rescher explains this reasoning: "With catastrophes we give way to the cautious pessimism of 'worst possible case' thinking. Here we do not calculate risks; we do not weigh probabilities; we simply note that there is a real chance of an absolutely disastrous outcome and turn to other alternatives, however unappealing in various ways."[26]

This precautionary philosophy parallels what has been described as "maximin" reasoning: a decision theory rule that urges us to "maximize the minimum." This strategy is of choosing the alternative with the best of all the possible worst-case results. This will never maximize benefits, but it will ensure that we never face the worst possible catastrophe. While the goal of expected utility is to maximize benefits, this rule advocates that we act with utmost caution, maximize safety and accept the opportunity costs in order to avoid the worst-case result.

Of course, as explained above, this strategy forgoes the possible benefits involved with taking risks, and we will never know what would have happened had we taken some other action, some other road. We might never know for sure whether our precautionary action is meeting a genuine need or whether it is but vain effort. The investment in possible catastrophe avoidance may make chumps of us all.

Not only might this investment be unnecessary, it would be difficult to implement. Because the cost is to individuals here and now, and the potential benefits are shared in common (we are discussing global safety here), prisoner's-dilemma type problems occur.[27] Strategists known as game theoreticians have identified certain situations, which they call prisoner's dilemmas, where co-operative behaviour is difficult to achieve even when the participants are aware that such co-operative behaviour will have more favourable results for themselves than non-cooperative behaviour. Prisoner's dilemma problems occur because there is not much point to co-operation unless we can be assured that most others will co-operate as well. So, for example, there is not much point in our taking steps to prevent global warming unless others do so as well. There would be no significant change in greenhouse gas emissions if a few thousand people, or even the populations of a few countries, stop driving cars, while the rest of the world continues to increase the rate of emissions. This problem is the current reason given by the United States for its resistance to co-operate in international agreements (such as the Kyoto agreement) on the reduction of greenhouse gas emissions.

The voluntary nature of present environmental agreements makes co-operation difficult to secure, and the lack of this security makes individuals and governments hesitant to act because nobody wants to undertake a plan of action that involves sacrifice unless they believe it will have a good chance of being effective. Nonetheless, cautious people believe that to be a chump is more acceptable than making mistakes in the other direction. I will defend this conclusion in chapter 10 on both moral and prudential grounds. With the costs of delays escalating, proponents of precaution have even rejected outright the necessity for scientific certainty: "If empirical proof is possible only when the biosphere is so overstrained as to reach the point of no return, it is prudent not to seek a scientifically satisfactory reply to the question."[28]

The precautionary principle is a strong one. It demands acting, even under conditions of scientific uncertainty, in anticipation of harm. It is not merely a principle for the prevention of known dangers. The prevention of further environmental harm from a known, harmful practice might be prudent, but this is not as strong an action as precaution. The precautionary principle states that scientific proof of a crisis is not needed to justify preventative action. Thus, the scope for precaution is much wider than for prevention, and this is why it is so controversial.

This principle is contrary to the epistemic conservatism that Bankes, above, claims has traditionally guided environmental decisions. Nonetheless, a general, but noncommittal, statement of the precautionary principle is often now included in the preamble of many international agreements, though not in Canada's domestic policies.[29]

The International Climate Change Convention of 1992 states: "Where there are threats of serious or irreversible change, lack of full scientific certainty should not be used as a reason for postponing such [precautionary] measures."[30] The 1992 Bergen Declaration on toxic substance control uses almost identical wording.[31] A 1987 North Sea meeting advised that "A precautionary approach is necessary which may require action to control inputs even before a causal link has been established by absolutely clear scientific evidence."[32] The 1990 Hague declaration on protection of the North Sea defined the principle as taking "action to avoid potentially damaging inputs of substances that are persistent, toxic, and liable to bioaccumulate even when there is no scientific evidence to prove a causal link between emissions and effects."[33] The Canadian Environmental Protection Act does not include a direct reference to the precautionary principle because it "is still rather vague and elusive."[34] However, the 1990 Canadian Green Plan did include the vague claim that we "should err on the side of protecting the environment."[35] The principle can be extended to apply to all aspects of human impacts on the environment. There are at least twelve different expressions of the precautionary principle in international treaties and declarations, including the 1987 Montreal Protocol on the ozone layer, the 1992 Convention on Biological Diversity, the 1992 Climate Change Convention, the 1992 Maastricht Treaty of the European Union, the 1992 Convention for Protection of the Marine Environment of the NE Atlantic, the 1992 Rio Declaration on Environment and Development, and the report of the International Joint Committee for the Great Lakes.[36] This is a principle that is taken seriously, and not merely a suggestion floated for further study.

The first formulations of the precautionary principle arose in Germany. The 1976 *vorsorgeprinzip* (the foresight principle, or the prudential principle) elevates the value placed on protection of the environment and prescribes avoidance or reduction of risks. It was used in Germany to justify protective policy measures with respect to acid rain, global warming, and pollution of the North Sea.[37] In Britain, early versions focused on the cost-effectiveness of pollution-control technology[38] and eventually developed into statements that parallel the above wording of the Climate Change Convention. A 1990 Government White Paper on the Environment, for example, states that "Where there are significant risks to the environment, the Government will be prepared to take precautionary action ... even where scientific knowledge is not conclusive."[39]

In sum, the precautionary principle forces a reversal of the traditional burden of proof. In some cases, it seems, the policy of demanding rigorous, scientific proof for models that predict harm is given up in favour of its reverse. The predictions are given the benefit of the doubt, and the burden of proof is transferred to those who wish to prove them false or unwarranted. According to strong versions of the precautionary principle, actions suspected of being harmful are assumed to be harmful until there is convincing evidence that shows otherwise. This policy has long been applied with respect to new drugs and food additives, for example. Anyone wishing to introduce a new food additive must conduct experiments to establish its safety.

PROBLEMS WITH THE PRECAUTIONARY PRINCIPLE

Just what the precautionary principle amounts to in practical terms is still unclear. Hill claims that a 1993 UK precautionary approach to genetic engineering has changed nothing except to introduce the need to ask for permission (which she implies is rubber-stamped).[40] Governments in the international arena might pay lip-service to the precautionary principle, but its actual force appears limited. This might be because of the enormous costs involved in precautionary action, as well as the vagueness of the principle. These problems create a problem of credibility for the precautionary principle.

Costs of the Precautionary Principle

The precautionary principle seems so commonsensical to so many people that one wonders why it was so long in coming. R.C. Earll, an environmentalist, argues for this commonsense point of view. "If you have little information about potential risks you exercise more caution, not less."[41] What could be more reasonable than anticipating harm and revising harmful practices *before* the damage is done? The conclusion that one ought to exercise maximum caution when so much is at stake is reached so often, so quickly, and so often without argument, that it is obvious that the often high costs of this caution are not fully appreciated.

The costs of precautionary action for environmental zero-infinity dilemmas are higher than one might first suspect. While estimates of the financial cost of reducing global emissions of greenhouse gases vary, even the lowest estimates are huge. The use of fossil fuels and the production of methane through agriculture are "economically fundamental,"[42] and greenhouse gases are a "byproduct of civilization."[43] To stabilize the atmospheric concentrations of CFCs, carbon dioxide, and nitrous oxide would require a 60 percent reduction in emissions.[44] The annual cost of stabilizing carbon dioxide emissions has been estimated at approximately 1–3 percent of the gross domestic product of the entire world (roughly equivalent to the entire output of Canada).[45] W. Steger estimates that carbon dioxide emissions reduction policies would cost the jobs of "at least six hundred thousand workers in America's most basic industries" and cause bankruptcies and economic inflation.[46] An estimate by W.D. Nordhaus, this time for a 50 percent reduction in all greenhouse gas emissions, is closer to 1 percent of the world output – still a staggering cost, roughly equivalent to approximately US$200 billion per year.[47] Of course the costs are not likely to be shared equally. Because each nation has different energy needs and resources, the costs of holding emissions to 1990 levels is expected to range from about 1 percent in the US up to a possible 13 percent in China.[48] A 20 percent reduction of carbon dioxide in the US alone could cost 3 percent of that country's gross national product each year.[49] These costs would be ongoing and are not just a one-time investment. Even some of those who favour precaution recognize the costs. Schneider writes that to avoid global warming "the developed world might have to invest hundreds of billions of dollars every year for many decades."[50] The diversion of such massive resources into projects meant to alleviate hypothetical problems means those resources are not available to solve more immediate and concrete problems like unemployment. It also means that less financial capital will be passed on to future generations.

Other precautionary costs might include drastically lower consumption rates, more expensive technology, less logging, or increased financial assistance to developing countries to ease technology replacement costs. In addition to financial costs, precaution demands that population growth be stopped. Practically speaking, it is difficult to see how this could be accomplished quickly without governmental interference in personal liberties. The precautionary principle might be taken to justify other interferences in our social lives as well.[51] P.J. Michaels, a libertarian, writes that even reducing greenhouse gas emissions by any significant amount would require "levels of command, control and tax coercion that will be abhorrent to most Americans."[52]

In addition to direct costs there are opportunity costs. The problem with refusing to take a certain risk is that you must also give up the potential benefits of taking that risk. Those who emphasize the small possibility of negative consequences of risk-taking often ignore the high probability of advantageous consequences. If the precautionary principle were ever taken literally, many slightly risky activities would have to be avoided, at great cost. While there may be risks in pursuing our present activities, we must also acknowledge the costs of avoiding these activities. Large areas of research in genetic engineering would have to be abandoned, along with the hope of enormously beneficial consequences for agriculture and human health. The emission of many substances would be banned altogether, millions of dollars would be spent detecting asteroids and developing rockets to eliminate them. We would have to restructure our society, one currently designed around the availability of cheap and plentiful energy. The stifling of research and industrial capacity, or the restructuring of agricultural production, might check the improvements being made in life expectancy and the capacity for food production. The cautioners are assuming that, if we avoid the risky activity, then at least we will be no worse off, but this assumption is not necessarily correct. Precautionary activity involves the loss of something that we might have had as well as the direct costs. (The details concerning the above costs will be disputed, of course, but I do not think it can be denied that there will be significant costs to precautionary action. If there were not, there would not be much need to debate the desirability of precautionary action, we would already be doing it. In chapter 10, I will argue in favour of precautionary action despite significant costs. The reader might be anticipating some of the arguments here – that externality costs of some activities have been overlooked, or environmental values have been underestimated.)

Of course, some precautionary actions might be relatively painless. If so, we have not much to lose by taking such actions – even if they do not, in the end, serve to alleviate any problem. For example, many governments deemed the possibility of ozone depletion and its negative consequences sufficient reason to give up aerosol spray cans. This action alone could not prevent the envisaged disaster, but it might lessen it somewhat, and it did not cost much either. For these uses, CFCs were easily replaced. One author estimates that a reduction of 10 percent to 30 percent of carbon emissions is possible at no cost.[53]

The recommended precautionary action is even more attractive if it actually pays off for reasons independent of the suspected risk. There are some minimal, protective actions that we can take that are dominant alternatives: the best course of action no matter what happens. For example, a policy of energy conservation has merits in its own right, regardless of whether this will also alleviate global warming. Energy efficiency has immediate pay-offs in terms of decreasing air pollution, and economic savings are usually achieved in fewer than three years.[54] Furthermore, reduction of fossil-fuel use might stimulate growth of other energy sectors or spur research in alternative transportation technology. The resulting decrease in carbon dioxide emissions might well do a great deal to alleviate global warming since carbon dioxide increase is the most significant factor in the greenhouse models. However, the point is that it does not matter whether global warming models turn out to be correct or not; energy efficiency is a good goal for other reasons. (The details of this particular example are hotly debated. Van Kooten argues that energy conservation is very expensive.[55] Blair and Ross argue that we can stabilize climate change and save money at the same time.[56] Whether the most efficient means of reaching energy conservation goals can be achieved through elimination of energy subsidies,[57] or the development of a new global market in emission shares,[58] is another problem. However, we are discussing the general notion of low cost precaution here, not this particular suggested action.)

An increased commitment to constructing and verifying models could also be included in this category of minimal cost precaution. Extensive monitoring and measurement of systems might reduce uncertainties in data for key areas of model input. Cloud and ocean studies might be conducted to test current mathematical parameterizations of complex processes.

If any actions can be found that fit this criterion of being both a protective measure against the risk of some global catastrophe and beneficial regardless of how that potential catastrophe unfolds, then of course these actions should be taken. However, most recommended actions for averting potential catastrophe are not so painless, and many argue that we simply cannot afford to spend money on unlimited precaution that might be unnecessary.

Vagueness of the Precautionary Principle

The precautionary principle has been very controversial because of its variety of definitions and vagueness of meaning.[59] On a generous reading, this vagueness is a reflection of the complexity of the problems and not a lack of intellectual rigour. While the general ideas of prudence, anticipatory care, and precaution are understood, the precautionary principle has been described in many ways and its details of are not well defined. Even though it has become a recurrent feature of international agreements, its meaning is open to interpretation. It is not clear, for instance, whether its central concern is with the likelihood of harm or the irreversible nature of the damage. Acceptable levels of statistical confidence have yet to be agreed upon, and these are needed to decide which risks fall into the domain of the principle.[60] In the last decade, the cities of Halifax and Dartmouth, in Nova Scotia, agreed that coastal regions of the marine environment have no assimilative capacity for absorbing sewage and other wastes, so dumping of wastes is now merely shifted to deeper water, where we do not know so much about the assimilative capacity.

The precautionary principle is also often set in the context of many other principles, such as principles of cost efficiency, that are not necessarily compatible with it and with which it must then compete.[61] However, there is no metaprinciple for deciding which of these competing principles takes priority.

Strength of the Precautionary Principle

In its weak versions, the principle makes precautionary action dependent on cost-effectiveness, competing interests, and estimates of the likelihood of the risk. So construed, its original aim of providing a check to the reckless destruction of the environment, excused by the lack of absolute certainty about the harm committed, is hampered. Without a strong statement of the priority of precaution, the weak version is not much different than cost-benefit analysis. Stronger versions of the precautionary principle, however, set a standard that is too high to live up to. The strong precautionary principle is too demanding because there is no end to the precautionary action we might take.

In its weak version, the precautionary principle is transformed into a modified cost-benefit analysis calculation with some added weight given to the cautious side of the equation. A zero tolerance for risk is dismissed as unworkable, and some environmental protection is sacrificed for economic and social benefits. However, the benefits of caution are assigned a larger value than would otherwise be assigned in order to provide a larger safety margin. The wording of these weak versions of the principle makes reference to "cost-effectiveness" and "best available technologies" rather than absolute prohibitions. Stebbing, for instance, argues that it is unnecessary to aim for a goal of zero-discharge for marine pollutants. The ocean can cope with a lot of chemicals without unacceptable damage, he claims.[62]

A satisficing decision strategy, as described in chapter 7, might be applied here. The point is simply to aim for a *satisfactory* outcome rather than the best possible outcome. What is needed is a basis for action without the need for absolute proof. Perhaps if the normally assumed burden of proof were reversed, so that what was required was "reasonable doubt" about the safety of various practices rather than proof that they were harmful, objectors would be somewhat appeased. The burden of proof need not be met always with conclusive proof, but merely with a defence against opposing arguments.[63] If we can somehow agree upon a minimum satisfactory outcome, then we can choose the option that guarantees it. Such a strategy will maximize neither benefits nor caution, but will at least ensure satisfactory outcomes. The 1992 UN Convention on Climate Change attempts this compromise between caution and development and seeks out some level of emissions that can be achieved without heavy cost.

Of course, determining a "satisfactory" emission level is hampered by the same problems we have been discussing. International agreements often mention targets, such as maintaining emissions at today's levels, or, for the 1997 Kyoto Convention of Climate Change, returning to earlier levels of emissions (say, those of 1990). The most recent IPCC report on climate notes that "stabilisation of atmospheric CO_2 concentrations at 450, 650 or 1,000 ppm would require global anthropogenic CO_2 emissions to drop below 1990 levels, within a few decades, about a century, or about two centuries, respectively, and to continue to decrease thereafter."[64] Bankes, however, thinks these levels can never be more than arbitrary choices.[65] The agreement reached on emissions reduction is a "soft commitment," writes Bankes, and applies to developed countries only.[66] Therefore, it might not go a long way in reducing the probability of global warming, and we might be making sacrifices that will not accomplish much. Michaels claims that the policy of stabilizing emissions at 1990 levels would make an insignificant difference to global climate. He estimates that by the year 2030 the global mean surface temperature of the planet would be approximately 0.25°C less than if no reduction takes place.[67]

The strong version of the precautionary principle, however, might be impossible to live up to. R.D. North warns that the precautionary principle is often stated in a way that would "outlaw all human action."[68] Caution is sometimes excessive and irrational. The precautionary strategy seems to recommend avoiding any activity that might possibly result in harm, and this strict avoidance of catastrophe would lead to almost total inaction. It is sometimes interpreted as requiring the elimination of all chemical emissions into the ocean, for instance, regardless of cost. A zero-emissions target is a goal of Greenpeace, for one.[69] "No wastes should be discharged in the sea unless it can be shown that they are harmless."[70] This, Greenpeace argues, is a rational judgment based on the insufficiency of data.

Consider also the moral dilemma of whether foreign aid to feed the starving will help to alleviate the problem or worsen it. The strict cautionary might resist helping because there is a small probability that aid merely exacerbates the problem.

The strict form of the precautionary principle demands an all-or-nothing stance. Because of uncertainty, it is not known how much or how little precaution is necessary. A strong precautionary principle demands the maximum caution. As an example of extreme (perhaps excessive?) caution, both A. Milne[71] and J.S. Gray[72] describe a debate in Norway over the disposal of large quantities of the mineral ilmenite mining tailings (a source of titanium). The Marine Research Institute of Norway recommends the construction of a very expensive dam to prevent discharge to the sea because ilmenite cannot be proved harmless. (Ilmenite causes the gills of crabs and prawns to blacken.) Because of the precautionary principle's reversal of the onus of proof, precaution is urged. This is an unreasonable precaution, according to Milne and Gray, because the mineral is common in many natural environments. Of course, it is possible that large quantities of ilmenite are harmful to aquatic organisms, even if it is found naturally. However, Milne and Gray adopt the stance of epistemic and practical conservatism and argue that it is unreasonable to protect against the mere possibility of harm that has not been shown to be statistically significant.

While the completely firm proof of harmlessness is logically impossible, so too is completely firm proof of harm. If a test fails to find any harmful impacts, that test has not proven conclusively that there are no harmful impacts. The test may be inadequate. Conversely, if a harmful effect is indicated by some test, there is always logical room for doubt about whether the correct or exclusive cause has been identified. Tobacco companies, for example, can still claim to doubt the causal connection between smoking and lung cancer because they claim there might be several causes, from genetics to diet. The two directions for the burden of proof are logically symmetrical, and it might seem that Milne's criticisms of the argument for the reversal of the burden of proof are too harsh.

What accounts for the traditional asymmetrical preference in questions of burden of proof then? It might be the scientific maxim known as "Occam's Razor," which advises that the simplest theory is to be preferred over one that postulates more new entities or effects than necessary to explain some phenomenon. The strong precautionary principle's demand for a reversal of proof allows that a hypothesis of harm must be accepted unless proven false. The traditional epistemically conservative approach is to reject hypotheses of harm unless proven true. In the context of environmental problems, conservatism assumes that there is no problem until proven otherwise. Precautionists prefer to assume that the environment is stressed or fragile, and that what needs to be proven is that it is not. Unfortunately, the precautionary principle allows for the multiplication of untold mechanisms of harm. (Of course more is at stake in some decisions, and this will be discussed in the final chapter. Right now, we are discussing the logic of the burden of proof.)

To make caution a useful strategy, we need some method of determining when taking caution is excessive. Rescher has recommended that we reduce the chance of excessive caution by ignoring very small probabilities.[73] When reasoning practically, "sufficiently remote possibilities," he writes, must be assumed to be zero. He argues that a decision must be made to dismiss risks with a probability below an agreed minimum. Thus, according to Rescher,[74] the best risk strategy can be described by three lexically ordered rules:

1 dismiss extremely remote possibilities
2 avoid catastrophes
3 maximize expected utility.

This strategy, however, is ambiguous when it comes to deciding just when to dismiss the chance of catastrophe. Once again, there is a question of where the burden of proof lies. Does it rest with those who wish to dismiss the risk as negligible? Or does the burden of proof rest on those who urge precaution to show that there is a significant chance of catastrophe?

Rescher argues that the determination of a threshold, beneath which risks will be dismissed as unrealistic, must remain a matter of subjective judgment. Subjective value judgments will be required in almost every stage of risk-taking decisions, including deciding upon what constitutes a catastrophe and what strategy to follow under uncertainty. Regardless of difficulty, the need to curtail unlimited caution is inescapable.

Is the Precautionary Principle Unscientific?

The precautionary principle has also been attacked for being unscientific. Milne argues that an environmental policy that advises taking action before convincing evidence is in, before causal links are discovered, undermines scientific practice.[75] The principle undermines science because it relaxes epistemic standards and because crying wolf will result in an eventual loss of authority for scientific judgments about the environment. The precautionary principle abandons the highest scientific demand, the need for incontrovertible evidence, and, paradoxically, some versions simultaneously demand evidence of safety by reversing the burden of proof. The precautionary principle abandons normal scientific standards of evidence without replacing it with some other systematic procedure for deciding exactly what to look for when estimating risks and harms. Milne accuses advocates of the principle of precaution of being unscientific and of practicing religion or moral philosophy instead.[76] The growing acceptance of the principle is sad news, says Milne. Similarly, Gray argues that the precautionary principle has "nothing to do with science" and that environmental decisions should be made using scientific evidence rather than rest on "reasons to assume" possible harm.[77] Gray thinks the precautionary principle, which was originally developed for deciding cases involving "persistent, toxic and bioaccumulatable chemicals," is now being extended, inappropriately, to other situations. He argues, further, that many current uses of the precautionary principle lack the objectivity of statistical tests.[78] Instead of demanding 95 percent certainty of harm, the precautionary principle settles for apparent trends. However, the existence of a trend in data is insufficient to justify warnings about negative environmental impacts. The "correct scientific approach," he argues, is to base environmental decisions on traditional statistical arguments.[79] While he claims that he actually supports the philosophy behind the precautionary principle, Gray disapproves of its "subjective applications," which "get carried away ... on unnecessary precaution."[80]

Milne, too, insists that the precautionary principle is useless as a legislative principle because of its impracticality. The principle, he claims, urges caution in order to avoid "harms" without requiring scientific evidence that is needed to identify harms in the first place. Furthermore, it cannot specify how much precaution is necessary. It is a common practice to estimate the assimilative capacity of an environment to absorb pollutants or stresses without long-term harmful consequences. Milne argues that the precautionary stance ignores this capacity in its demand for ultimate caution.

PRECAUTIONARY PRINCIPLE VS PRACTICAL CONSERVATISM?

This chapter outlined two possible strategies in response to the possibility of global catastrophe. Epistemic and practical conservatism is the strategy of not reacting to suggestions of catastrophe unless there is sufficient proof that one has an unacceptable probability of it occurring. Unfortunately, this conservatism in environmental policy-making might lead to catastrophe. The precautionary principle, on the other hand, urges preventative action without the need for scientific proof of impending catastrophe. We have seen that there are serious problems with the precautionary principle. In fact, one might even think these problems are insurmountable. However, we will see in the final chapter that there are possible defences of precaution. In the next chapter, we will discuss a different approach. The decision dilemma described here has led some to call for new conceptions of rationality.

·9·

Rival Rationalities: Democratic Epistemology

We saw in the previous chapter that, on the one hand, the precautionary principle has seemingly intractable problems and that, on the other, epistemic and practical conservatism might lead to global catastrophe because it encourages inaction in the face of enormous risk. In chapter 10, I will come back to the problems with the precautionary principle and offer a defence of it. First, though, I want to discuss another way of responding to the lack of definite probabilities for predictions of global risk. Some authors argue that, if normal accounts of rationality do not give us reasons to act precautiously, then we ought to adopt an account of rationality that does. We have seen from the discussion of decision theory that epistemic values must compete with other values in the process of reaching decisions about what to do. I will illustrate how the competition might be operating for cases of decisions about global risk. Some want to make a stronger claim than this, however. They want to argue that what we should *believe* should be determined by what we are most committed to *do*. I will critique this view and claim that it is a result of running together epistemic values and action-guiding values. We must distinguish carefully between claims that something is the case or is probably the case, and claims that some course of action should be pursued.

Some ideas are associated so often that we often forget that they can be separated. In the exercise room at my university, for example, the idea of exercise has been inextricably tied up with the idea of playing loud music. When the varsity basketball teams practice in the adjacent gym, or when exams are written there, the exercise room has always been shut down so that the loud music would not interfere with these other activities. It is only very recently that the people in charge of the athletic centre have figured out that all that needs to be closed down is the sound system–and that it is possible to exercise in silence.

Two ideas in environmental decision-making have also been so long associated that many people forget that these ideas are, in fact, separable. People often link ideas about what to believe with ideas about what to do. It is accepted by many partisans on both sides of the debate about global warming, for example, that if we do not have good reasons to believe in the accuracy of global warming predictions then, correspondingly, we have no good reasons to change our activities in ways that would reduce greenhouse gas emissions. Conversely, if we have moral or political arguments that would encourage us to change our ways in the face of an environmental threat, then some people feel compelled to conclude that we have reason to think the science predicting catastrophe is right on target. These people argue that we should adopt an epistemology that places more emphasis on democratic and ethical values. They often point out that decision-making about global risks has been mostly a matter of ideology anyway, so we may as well openly discuss these values and make decisions for political, cultural, aesthetic and ethical reasons rather than on the pretense of scientific grounds. It will be my concern, in this chapter and the next, to separate these two questions of action and belief and to argue that one can have good reasons for precautionary action with respect to environmental risks despite the lack of epistemological certainty about the predictions of harm. There are good reasons for adopting a precautionary program even if, along with the risk-takers, one strongly doubts that our present ways will in fact land us in hot water. I will compare some environmental decisions to Pascalian wagers and argue that one need not believe that the evidence for global warming is convincing, for example, in order to act as if were.

Those who make the mistake of running together questions of epistemology and questions of what to do stand in good company. James Wernham points out that Pascal and William James have made the same mistake in their famous arguments on the belief in God. In chapter 8, I described Pascal's argument for the belief in God. Pascal argued that one should believe in God because the stakes involved are so great. If God exists, then the nonbeliever has lost something of infinite value – a heavenly reward of eternal bliss. On the other hand, if it turns out that God does not exist, the believer has not lost much. Wernham, however, argues that despite the usual interpretation of Pascal's argument, it is not about believing, but about gambling.[1] When someone makes the wager in favour of God's existence, that person is acting – not believing. One need not believe in God to gamble that God exists. What one stakes in betting that God exists is one's old way of life – not an epistemic belief.

William James, too, fails to distinguish action from belief. James's famous essay "The Will to Believe" is usually interpreted as an argument to justify a belief in God. James argues that, under certain conditions (which I will explain more fully in chapter 10), it is not irrational to form beliefs even if there is insufficient evidence for those beliefs. Belief in God is one example of a belief that one is justified in holding despite the lack of certainty. Wernham proposes that, while James's essay is usually interpreted as an argument about the ethics of belief or the duties of inquiry, and this is the way James himself describes it, it is better reconstructed as an argument for the prudence of acting one way rather than another. Thus, it is not really about either ethics or belief. Rather, it is about wisdom in gambling, gambling understood as performing an action. James's argument is often mistakenly interpreted as concerning ethics and belief, Wernham says, because James took himself to be responding to the arguments of Huxley and Clifford – arguments about the ethics of belief. James's argument makes more sense, however, if seen as a Pascalian argument about wagering. On Wernham's reconstruction, James's argument supports only the claim that the prudent course is to act as if God existed and does not support any claims about belief.

In this chapter, I will argue that one strand of epistemology might be making the same mistake of linking questions of epistemology with questions of action. This strand argues that, if normal accounts of rationality do not give us reasons to act precautiously, then we ought to adopt an account of rationality that does. These authors make the strong claim that what we should *believe* should be determined by what we are most committed to *do*. I will critique this view and reject both democratic epistemology and procedural rationality for their failure to distinguish properly between epistemic values and action-guiding values. We must distinguish carefully between claims that something is the case or is probably the case and claims that some course of action should be pursued. Some have made the mistake of thinking that one must believe in predictions of environmental disaster in order to act on them. However, environmental decisions are no different from Pascal's wager when it comes to the absence of the need for belief. There is no need to believe that models of environmental systems predicting harm are true in order to bet on them. Acting precautiously in response to some model that is predicting environmental harm is a gamble, not a belief in the truth of that model.

This running together of questions of belief with questions of action is a result of trying to explain the role of values in epistemology. It is obvious to many observers that values seem to affect what people believe. I will illustrate below, for example, how those who do not believe in global warming are usually libertarians who hate paying the taxes that will likely come with reductions in greenhouse gas emissions; or they live in oil-rich places like Alberta; or their research is funded by the coal lobby. Similarly, those who are confident that data already reveal the existence of global warming are usually those who have vested interests in believing that to be the case.

Science is not supposed to work that way. It is supposedly an objective enterprise where the motives and values of the scientists should not affect the evaluation of theories. Only bad science, it is often thought, would result in the observations I just reported. Some theorists, known as "social constructivists," and some feminist epistemologists insist that this is a naive view of science and that values are inseparable from the process of justification and epistemology. Every belief serves somebody's interests, they point out, so we may as well believe that which serves democratic interests, or feminist interests. Helen Longino writes, for example, that "Instead of remaining passive with respect to the data and what the data suggest, we can acknowledge our ability to affect the course of knowledge and fashion or favor research programs that are consistent with the values and commitments we express in the rest of our lives."[2] I will argue that this move is unnecessary to solve the problems involved with the precautionary principle. We can defend the need for precautionary action on ethical grounds without requiring a complementary epistemological belief.

To facilitate the discussion of the role of values in environmental decision-making, I will make use of two examples. In the past, decisions about how to respond to possible global threats were made for reasons other than merely the strength of the scientific models involved. Policy-makers sometimes used the bare possibility of what they deemed undesirable outcomes to justify action meant to forestall those outcomes. In some cases, decisions about global risks have been made largely on the basis of political commitments and despite the absence of convincing evidence. Models found wanting scientifically were nevertheless accepted as providing sufficient grounds for decision-making. In hindsight, whether the resulting decisions turned out to be the correct ones or not has largely been a matter of luck.

The models both of ozone depletion and of nuclear winter were accepted in their early stages as revealing significant global risks. The general consensus now is that, in the ozone depletion case, fear of the risk turned out to be justified, while in the case of nuclear winter, faith in the model turned out to be unjustified. The effects of dust kicked up by a nuclear war would still be serious, but it is now generally thought that these effects would be much less serious than first portrayed in nuclear winter models. In the case of neither of these models was there sufficient evidence for thinking that its predictions of catastrophe were reliable, yet people did accept them and thought that they provided sufficient reasons for action.

If the models' predictions were unsupported by convincing evidence, why were they nonetheless accepted? More than a few observers have suggested that they were accepted for political reasons.

Despite their weaknesses, nuclear winter models gained a great deal of support and were taken as supplying one more reason for limiting and reducing nuclear armaments. The rapid acceptance of the nuclear winter vision might have been a result of researchers' concerns to be politically correct. In "A Memoir of Nuclear Winter," Tony Rothman reports on his interviews with some major participants in the debate and concludes that this widespread acquiescence in the promotion of a suspect model was a result of the noncritical stance that most scientists adopted for political reasons.[3] Rothman argues that the theory of nuclear winter gained wide support because of the politics involved rather than the science behind the theory. Specifically, nuclear winter models were quickly adopted because they gave support to a policy of weapons reduction. Rothman describes how many scientists uncritically supported these weak models because they thought that it would be immoral not to support a theory that furthered the political aim of nuclear disarmament. Rothman quotes Freeman Dyson's reaction to the theory: "My instincts as a scientist come into sharp conflict with my instincts as a human being. As a scientist I judge the nuclear winter theory to be a sloppy piece of work, full of gaps and unjustified assumptions. As a human being ... nuclear winter is not just a theory – it is also a political issue with profound moral implications."[4] Rothman laments that current, better models of nuclear winter would have more influence on policy today had more moderate conclusions been publicized in the first place, rather than the apocalyptic versions that later had to be withdrawn – adding to the perception that models of this kind are cries of wolf.

Models of ozone layer depletion were also able to generate confidence in the predictions of catastrophe despite the absence of direct evidence that CFCs were actually causing any damage. The confidence in the models was sufficient to initiate worldwide preventative action *before* there was any measurable evidence.[5] Action was taken to reduce the use and release of CFCs as early as 1978. However, it was not until 1988 that a study of the Antarctic atmosphere was published that revealed that there was a direct correlation between increased concentrations of chlorine and decreases in ozone.[6] In *Beyond the Limits*,[7] Meadows et al. speculate that the decision to reduce CFC use was made because of the general climate of environmentalism.

Parallelling the nuclear winter case, the model was accepted before the evidence was convincing. Also analogously with the case for the nuclear winter model, the model's predictions were accepted for political reasons rather than on the model's scientific merits.

Supporters were proved correct about the existence of ozone depletion. (It appears now that they had even vastly underestimated the magnitude of the problem.) In the nuclear winter case, however, the models portrayed a catastrophe that is now believed to be exaggerated. How do we know which error we are likely to commit next time? Models of global warming and models of overpopulation and limits to growth are also unable to provide convincing evidence of the likelihood of catastrophe. How, then, should we react to them? Should we follow the example of the ozone layer depletion case, be cautious and act on the assumption that predictions of catastrophe will turn out to be correct? In this case, policy-makers took early action that later proved warranted. Or has the nuclear winter example taught us not to respond to cries of wolf? After all, there is almost no end to speculations about possible catastrophic futures.[8] Should we assume them all to be true and take the immediate precautionary actions that supporters of the models of global catastrophe say are necessary? Or should we remain skeptical, and take minimal action, or no action at all? How are current dilemmas about global risk being decided, and to what extent should they be based on purely scientific considerations?

Given the state of uncertainty that surrounds almost every aspect of the greenhouse models, what should we believe about the enhanced warming predictions? S.A. Boehmer-Christiansen supplies reasons for believing that models of enhanced global warming have gained much more political purchase than the ambiguous science warrants because of the current geopolitics of energy.[9] She argues that it is no coincidence that the global warming hypothesis gained widespread acceptance in the late 1980s even though it had been proposed decades earlier. Models of global warming have gained their favourable status, she thinks, because of the current politics of energy, which are the result of the precipitous rise and fall of energy prices in the 1980s. She writes that "Evidence ... suggests that the proactive 'global warming' alliance succeeded during the mid-1980s because the epistemic community had been able to attract powerful friends in energy institutions which desired state intervention in energy markets."[10] The greenhouse models are given more credit than the science warrants, she thinks, because the threat of global warming is being used by influential persons as the "justification for a crusade against materialism," or for rationalizing energy policies "as environmental even though adopted for other reasons," reasons such as the promotion of nuclear energy and new fast breeder technology. Various other agendas were also promoted: "separate groups pursuing commercial interests, foreign policy goals and domestic politics each discovered their own uses for the warming hypothesis."[11] She claims that the 1990 IPCC assessment itself reflects the politicization of the scientific debate and was produced to legitimate agendas that were already decided.[12] Boehmer-Christiansen concludes that models of global warming are a "skilful exercise in scientific ambiguity" that have been used as tools to justify contradictory policies. Therefore, she says, their relevance to policy is in question. Those with vested interests in carbon energy will oppose any carbon dioxide reduction policy, while those who stand to gain from such a policy will support it.[13] The uncertain models have become a superfluous cog in the middle. She concludes that the models of global warming are still too weak and that their claim to policy relevance should be questioned because the uncertainty is used to advance unrelated agendas.[14]

Others, too, argue that we have already given the models more attention than they deserve. P.J. Michaels notes how the "global warming thesis has swept public and political opinion despite its many weaknesses,"[15] and Wildavsky argues that the greenhouse debate simply provides a new backdrop for the same old ideological debate between regulation versus free enterprise – a debate whose engine is fuelled by various political and economic motives rather than belief in the accuracy of the models.[16]

Proponents for and against risk-aversion might only be reflecting their own personal value set. Libertarians might argue against precaution because they do not want their taxes to go up. The green pessimists might really be motivated by a self-righteous moral program that condemns our current emphasis on economic growth as morally reprobate, critics like H.S.D. Cole suggest. He writes: "The flavour of environmental critiques is one of utopian preference for ascetic and egalitarian poverty over advanced technology as a means to abundance."[17] The precautionary principle is a challenge to the theoretical conservatism of standard scientific reasoning and the business-as-usual industrialists and thus, it is argued, attracts a certain type of personality. The precautionary principle has survived despite its vague, nonoperational definitions because it captures "emotions of misgiving and guilt."[18] The criticism being advanced here is that the values of environmentalists cause them to overstate the probability of catastrophe. Underlying values will cause cautious people to emphasize overpopulation and global pollution as likely to bring about catastrophes, while others emphasize the missed opportunities, the failure to have a prosperous economy, and the loss of potential families.

Hans Jonas and others argue that using doomsday scenarios as the fastest path to obtaining some ethical goals should be encouraged. Jonas argues that the presentation of possible catastrophic scenarios might be enough to scare us into acknowledging the value of what we now take for granted.[19] A Joni Mitchell song reminds us: "Don't it always seem to go / That you don't know what you've got / Till it's gone."[20] Perhaps we need to be able to imagine doomsday before we will take steps to prevent it. Thompson and Schneider have argued, for instance, that despite the weakness of nuclear winter models, the fear they generated spurred political solutions to problems in the arms race.[21] Jonas feels that it is the "first duty" of an ethics of the future to imagine and model worst-case scenarios in order that we fear them enough to avoid them: "We need to seek out evil so that it can instill in us the fear whose guidance we need."[22] Since it is human nature to avoid painful decisions until we are forced to make them – that is, until we are already suffering from being hit over the head with the results of our avoidance of an unpleasant truth – Jonas suggests that we should do everything possible to make the worst-case possibility as vivid a possibility as can be. The probabilities may be unknown, but we can conceive what the range of outcomes may be by modelling possible outcomes. These "heuristics of fear" may help us realize what we value and want to maximize in our decisions. Jonas argues "As long as the danger is unknown, we do not know what to preserve and why. Knowledge of this comes, against all logic and method, from the perception of what to *avoid* … *We know the thing at stake only when we know that it is at stake* … We know much sooner what we do not want than what we want. Therefore, moral philosophy must consult our fears prior to our wishes to learn what we really cherish"[23] (his emphasis). Jonas argues that we should encourage proliferation of predictions of catastrophe and present plausible doomsday scenarios to the general public in the hopes of creating a commitment from citizens to the survival of the human race. It is hoped that fear will stimulate government and public action that is normally slow to materialize unless there is conclusive evidence of catastrophe.

Similarly, Stephen Schneider, a climatologist for the National Center for Atmospheric Research, has acknowledged the temptation "to offer up scary scenarios, make simplified, dramatic statements, and make little mention of any doubts we might have … Each of us has to decide what the right balance is between being effective and being honest."[24] Thus, for political reasons, the truth might be sacrificed in order to achieve a desired political outcome.

That a noble lie might sometimes work to produce a desired outcome is an obvious truth. Motivational gurus tell us that if we can convince ourselves, for example, that everything bad that happens to us in our personal lives happens for some purpose, then we will probably lead more satisfying lives than if we believe in a random, meaningless universe in which bad things happen for no reason whatsoever. Similarly, a widespread fear of global disaster might lead to better, less greedy, less aggressive societies. The belief in the pressing need for changes in the way we move through the world might make these changes come true. The fear of the end of the world has been the spark of radical change before. Controversially, Albert Schweitzer argued that it was Jesus's mistaken belief that doomsday was imminent that structured his radical new ethics.

This strategy of noble lying, however, will not work in the long run. P.R. Gross and N. Levitt[25] and Rothman warn of the dangers of crying wolf. Not only is the move away from cautious scientific reasoning likely to result in policies of overreaction, they claim, but the fears of catastrophe will wear off if they become too commonplace.

A further problem is that political and ethical considerations may promote or smother an uncertain theory prematurely. We think it irresponsible to act as if science can be divorced from its political consequences, and this leads some to think it irresponsible to voice those scientific theories that are not commensurate with our prior political or ethical commitments. We might be reticent to criticize the theory of nuclear winter because we are afraid criticism will weaken political commitments to disarmament. One ecologist was severely criticized for publicizing his view that, according to his evidence, the problems of species extinction associated with deforestation have been exaggerated. The ecologist was accused "not only of bad science but of encouraging deforestation ... (and hence) of a serious breach of ethics."[26] We might be quick to promote theories of ecology that increase the fear of looming catastrophe and hesitant to promote those theories that might lessen this fear in order to make these theories reinforce our current political or ethical attitudes. Rothman concludes, from his observations on the nuclear winter debate, that "it is not clear ... at what point a scientific theory should become the basis of political issues, but my instincts tell me that the science should be established first, the politics second. It is ... dangerous to base the moral argument on a tentative scientific result."[27]

In a recent attack upon the idea that policy responses to threats of global catastrophe should be dictated by politics, Gross and Levitt accuse environmentalists of overstating the probability of catastrophe for ideological reasons. They accuse environmentalists of "ecotopian enthusiasms"[28] and of supporting apocalyptic models because of an ideology of puritanism and the search for compelling reasons to reconstruct society, rather than because of an "exactness of scientific thought."[29] Proper decision-making, they claim, requires accurate assessment of probabilities. On this view, the most relevant information for decision-makers is the scientific information, such as that provided by rigorous, plausible models. Science provides value-free evidence that must then be assessed and, through value judgments, transformed into policy. In the absence of strong evidence, we should suspend judgment, letting neither our hopes nor our fears hold sway. Boehmer-Christiansen claims that the International Panel on Climate Change has adopted this view, a view she describes as the "linear, apolitical model of the policy process."[30]

When the science is uncertain, however, Boehmer-Christiansen argues, there are severe limits on the ability of science to provide advice for policy-makers. We should not be looking to science for the primary justification of policy anyway, she claims, because this overlooks the many other interests at stake. The implication that science will decide environmental policy has allowed various governments and nongovernmental groups to use the science of climate change as an "empty justification" to advance predetermined agendas, such as urging the adoption of a new technology, rather than use the science to inform these agendas. Even the decision to do nothing but wait for the results of more research misleads us into thinking that some future scientific knowledge will decide the question. Governments that take this approach of waiting for more results have conceded that the models are the source of the relevant knowledge and thus have sought to abdicate responsibility. Postponing the decision, in the belief that scientific certainty is required for rational policy-making, is a way of allowing environmental policy to be decided by interests that remain unexamined, she warns.[31]

Kierkegaard explored the question of decision-making under uncertainty, and pointed out that our choices do not merely reflect who we are but construct who we are. In *Either/Or*, he makes an analogy between decisions made at the precipice at the end of rational knowledge and a ship that keeps making headway while we wait to decide.[32] If you do one thing or the other you might regret your decision either way, but, he argues, you have to decide who you are – the decision "constitutes our personality." Kierkegaard was discussing spiritual issues that cannot be resolved by better science. However, one could make a useful analogy with our decisions about global doomsday forecasts. These decisions determine who we are as a society and who we might have been. We must, insists Kierkegaard, accept responsibility for our actions, even though they can never rest on infallible reason and certainty. We must decide what course of action is the most reasonable and acceptable for our society and act accordingly. Applying Kierkegaard's ideas, we can say that our response to doomsday warnings will structure the kind of society we live in and determine who we are. Deciding to act now to take measures to limit risk (even if that risk turns out to be nonexistent) will define us in a different way than deciding to wait and see.

Mark Sagoff has taken this leap of faith. Along with Boehmer-Christiansen, he thinks that we should not be looking to science to inform policy in areas of great risk. Let's get straight to the values, he urges. Because the science of ecological modelling is so uncertain, he argues that we should let aesthetic, political, and moral arguments dominate the decision-making process.[33] If models of global systems are unable to provide trustworthy estimates of probability for global catastrophes, then, according to Sagoff, we should not give them much weight in policy-making.

What is going on here? The above debate seems to be about whether our decisions about what to do when faced with environmental problems should be decided by science or by political, ethical, aesthetic, and other considerations. Both sides use uncertainty to justify their claims. On one side, it is claimed that the models of global systems that predict catastrophe are too uncertain to justify taking precautionary action. On the other side, precaution is urged precisely because of this uncertainty. This debate is also confused with debates about what to believe about the probability of catastrophe. To sort this out, we must distinguish between belief and action, models and policies. There really cannot be a conflict between science and politics, for science can say nothing about what should be done and politics can say nothing about what is the case.

It should come as no surprise to us that values and other nonscientific matters are central to decision-making. As I explained in chapter 7, reasons for action are always belief-desire pairs. The resources of science can help us form our beliefs about what is the case, but decisions are not only scientific matters. They are also social, political, moral, and religious matters. In chapter 8, I described an argument by Milne and Gray, who both attacked the precautionary principle for being unscientific. The "correct scientific approach," Gray argues, is to base environmental decisions on traditional statistical arguments.[34] The accusation of practicing moral philosophy, which accompanies their argument, will come as no insult to many who think that practicing moral philosophy is appropriate in some cases and is nothing to be ashamed of. Milne's and Gray's criticisms can easily be deflected by taking the high ground. All one needs to do is simply acknowledge the nonscientific nature of the precautionary principle and argue that a policy principle, not a scientific one, is what is required here. The precautionary principle is a political response reflecting a perceived duty to protect the environment in cases of uncertainty. J. Lawrence and D. Taylor argue that scientific and political judgment have separate domains and the precautionary principle falls into the realm of political judgment.[35] That the precautionary principle is not scientific should also not be surprising to those who know the genesis of the precautionary principle. R.C. Earll reminds us that the precautionary principle was a response to past failures in risk-taking decisions and unacceptable pollution levels in the present – it did not develop as a scientific conclusion.[36] The precautionary principle is not scientific, nor is it anti-scientific. Rather, adopting the principle is a decision to *act* one way rather than another, to gamble this way rather than that.

To see why the precautionary principle is not a scientific principle, imagine how one could respond to the criticism that the precautionary principle does not really add anything to the debate because it is merely another way of stating the strength, or lack of strength, of scientific evidence. Even if the precautionary principle were accepted, and the burden of proof were reversed, the argument goes, this would not guarantee safety. Many of our current environmental problems are the result of pollutants that were once thought to be perfectly safe. CFCs, for example, are nontoxic, stable, and do not burn or react with other substances. Some activity might pass every safety test available and yet cause harm in some unknown, unexpected way. We cannot ban all activity because of ignorance. Thus, we apply the precautionary principle only in cases where there is strong suspicion of harm. However, these suspicions are only strong when the scientific evidence is strong. When the evidence is uncertain, the justification for precaution is also weak. Is it true that precaution is justified only in direct proportion to the strength of the scientific evidence? Well, no. Precaution is an attitude that can justifiably be adopted across the spectrum of strength of evidence. The precautionary attitude is dictated more by issues such as the catastrophic nature of the risk, whether the risk is shared equitably, and other nonscientific factors rather than by the strength of evidence.

Recall from chapter 7 Giere's description of decision theory and his recognition of values other than epistemic ones in the process of deciding what theory to adopt, or to work on. Extrascientific interests are included right from the start when assigning values to possible outcomes. When there is a lot of evidence about states of the world, when probabilities are known, then the epistemic values are strong enough to overwhelm the others. However, in cases of uncertainty, political, moral, religious, and other values will determine the ranking of various outcomes. His example was the resistance of geologists to plate-tectonic theory. In the absence of convincing evidence, the ideological commitments of geologists in the 1920s led to a rejection of the theory of continental drift. Giere argues that this was not a mistake of reasoning, but only illustrative of the role of nonepistemic values in decision-making. The plate-tectonic theory was eventually accepted as evidence mounted in its favour. Giere's example illustrates that while both facts and values are necessary to make decisions, the roles played by facts or values can be weak or strong. The more uncertainty increases, the more policy-makers have to rely on values alone. However, even when there is a clear scientific basis, values and politics are still relevant.

DEMOCRATIC EPISTEMOLOGY AND PROCEDURAL RATIONALITY

A. Seller, K.S. Shrader-Frechette, and E.D. McCoy argue for a different solution than Giere's to this debate about the roles of science and values in decision-making. Rather than keep separate scientific and political questions – questions about what to *believe* and questions about what to *do* – they want to blur the distinction. What we believe, they argue, should be conditioned by how we ought to act.

Shrader-Frechette claims that there is a middle way between the two biases of science and politics, a way that she calls "procedural rationality." She draws an analogy between debates about risk evaluation and a debate in the philosophy of science about what constitutes rational theory choice. On the one hand are those philosophers – for instance the logical positivists – who think that there are objective methods and rules for choosing one theory over another. On this account, all rational people should agree about which theory best explains the data. On the other hand, there are those who claim that there is no objective scientific method and that epistemic decisions are socially constructed. In the middle are those who believe that science is objective even though there is no scientific method that is free of value judgments. This is a debate about epistemology and what to believe. Shrader-Frechette notes the structural parallel between this debate and one about risk evaluation and policy formation.[37] The question here is whether there is an objective, rational method for determining an appropriate response to risks. As I have already mentioned, some think that our decisions about what to do should be primarily based on objective and scientific criteria – a position Shrader-Frechette refers to as "naive positivism." Others think that these decisions should be primarily based on political criteria. Shrader-Frechette argues that both these positions erroneously reduce risk assessment to either pure value judgment or pure scientific rules. A purely scientific approach to risk assessment ignores citizens' concerns about equity and consent, while a purely political risk assessment is too subjective and arbitrary.[38] She argues, instead, for what she calls a middle position. On her account, science and politics are merged. So too is the distinction between belief and action.

Anne Seller makes similar statements in arguing for what she calls "democratic epistemology." In cases of uncertainty, we allow "our political commitments to decide our view of the truth."[39]

Shrader-Frechette, McCoy, and Seller set out to defend an ethical basis for decision-making about what to do in cases of uncertainty by arguing that decisions in these cases are based on beliefs that are themselves laden with values. I will outline their positions in more detail and then argue that this decision not to distinguish clearly between what we ought to believe and how we ought to act is confused, unnecessary, and counterproductive. This confusion is unnecessary because Giere's satisficing decision theory described in chapter 7 will suffice to incorporate both science and values in decision-making without the need to redefine rationality. For Shrader-Frechette and McCoy to set up a dilemma between purely scientific decision-making and purely ethical decision-making might be, as Mary Douglas argues, to set up "artificial enemies."[40] The confusion between belief and action is counterproductive because, in calling for a radically new conception of rationality, Shrader-Frechette, McCoy, and Seller weaken their case for precaution by making it dependent on a suspect new conception of epistemology. I will argue in favour of a more traditional view, namely, that there is a distinction between claims that something is the case or is probably the case and claims that some course of action should be pursued.

Democratic Epistemology

Seller promotes an epistemology that merges the question of what to believe with the question of what to do. She argues against the traditional view that we can first decide what the truth is and then decide what to do about it. Under certain conditions, we "cannot prioritize the problems," she writes, because "each of these decisions affects the others."[41] Instead, we must decide, at the same time, both what the facts are and what to do. Epistemological questions are decided by an ethical or political procedure.

Seller would like to believe in what she calls a "rational-scientific" epistemology because this would allow feminists to defend against what she sees as sexist claims about, say, different kinds of intelligence between the sexes.[42] This, and other similar claims, could be discredited by appeal to factual, realist claims about truth. However, this search for objective truth, she fears, is betrayed by the interests of the searchers. Since beliefs always serve some particular interests, epistemologists should properly attach riders to factual claims explaining whose interests are being served by belief in those factual claims. The influence of interests over belief is most easily seen when there is a disagreement over what the facts are. As an example, Seller supposes that there is some question about the statistical likelihood of childhood leukaemia in the neighbourhoods of nuclear power stations. The data are not conclusive. While some surveys reveal that nearby residents do have a higher risk, others claim that the risk is the same as in the general population. Furthermore, she supposes that radiation is not a sufficient cause for leukaemia and that poor nutrition might also be a factor in the development of the disease. In such cases of uncertainty, Seller argues, we often "get very close to allowing our political commitments to decide our view of the truth."[43] Because of different background values, the managers of the nuclear power plant and residents might disagree, in this example, about whether radiation or poor nutrition is the cause of the disease and about whether there is an above average risk of leukaemia near nuclear power stations. Many people will base their beliefs about what is the case on their "prior commitment against nuclear energy."[44] This is not irrational, she claims, but an appeal to a different set of value judgments. Out of a network of causes for leukaemia, we focus on different contributing factors depending on our interests.

Knowledge, she continues, is like politics. Both are best seen as a "process, rather than as achievement."[45] She argues that both epistemological and political reasons for decisions "need to appeal to the same community, in the same way, in order to decide what is the case and what we should do about it."[46] This "democratic epistemology," as she calls it, is not to be considered second-rate because all reasoning, including scientific reasoning, is in the same boat.[47] There is no way to settle disputes and disagreements about any issue of knowledge except by consensus. Rival points of view compete in a procedural mechanism of open discussion, and the truth is whatever results from this procedure. Knowledge, she claims, is about control and solving problems, not about correct descriptions of the world.[48]

The reader might immediately object that this advocacy of democratic epistemology borders on relativism. Seller, however, denies this charge: "I am arguing that the ultimate test of the realist's views is their acceptance by a community, and that it is also in a community that the non-expert decides what to believe and what not. This is not to say that such acceptance makes those views true, or rejection makes them false, but rather to say these are the best methods available for making decisions which accord with reality."[49] It is Seller's contention that procedural accounts of rationality can still lay claim to objectivity because objectivity is also interpreted in a procedural way. The objectivity comes from the fact that the debate is about publicly observable events seen from a wide range of alternative prior commitments. This notion of objectivity is in the tradition of Wittgenstein and Rorty because it is "tied to the *practices* of people" rather than to mind-independent events.[50]

Applying Seller's arguments to our problem of what to believe about the predictions of models of global systems that predict catastrophe (unless we take preventative action), we see that assessments about the probability of risk will differ depending on prior political and other commitments. The final assessment will be determined by democratic procedures, however, and not by appeals to scientific objectivity that Seller thinks are elitist and authoritarian.

Procedural Rationality

Shrader-Frechette argues for a similar revision of epistemology. She advances the case for procedural rationality in *Risk and Rationality*, and later, along with co-author McCoy, in *Method in Ecology*. After raising the concern about how standard accounts of rationality favour suspended judgment in the face of uncertainty, Shrader-Frechette and Mc-Coy ask "whether scientific rationality provides the best model for environmental decision-making."[51] They conclude that it does not. Instead, they propose that we adopt a decision method they refer to variously as "ethical rationality," "ecological rationality," "procedural rationality," and "a negotiated account of risk and rationality."[52] Shrader-Frechette describes it as "a whole new theory of rationality, one that is critical, procedural, populist, egalitarian, and democratic, as well as objective, [and] scientific ... in other words, an epistemology in which *what* we ought to *believe* ... is bootstrapped onto *how* we ought to *act*"[53](her emphasis). Shrader-Frechette wants to develop an account of rationality that incorporates directly the ethical and political aspects of problems. On this account, belief is constrained by democratic principles. Since all science is value-laden, Shrader-Frechette and McCoy claim, why not incorporate democratic, egalitarian, and other values?

Their main argument stems from a view that all science is value-laden. They argue that, because the evidence for models or theories is always uncertain and underdetermined, researchers must "fill in the gaps" between hypotheses and evidence with "methodological value judgments."[54] Because the facts are never completely known, "personal, social, cultural, ethical, or philosophical" evaluative assumptions, they claim, supply the bridge enabling scientists to make inferences from incomplete empirical data.[55] Empirical data are meaningless without an interpretation, and interpretation, they argue, is necessarily laden with methodological value judgments. Different methodological value judgments will result in different interpretations of what the facts are. These evaluative assumptions are unavoidable *in principle*, because all research and investigation requires judgments about which scientific practice to follow. These judgments, they claim, are influenced by ethical commitments and are never purely factual.[56] Following Helen Longino, they argue that some of these value judgments are not simply reflections of bias, which might be eliminated, nor contextual values that serve only to guide the direction of research. Rather, they are inescapable "constitutive" values that determine the methods used in experiments. They are value judgments that transcend the traditional distinction between the context of discovery and the context of justification and are thus beyond rational criticism.[57]

Most everyone accepts that values and biases play a large part in science. However, it is usually thought that these values and biases are restricted to the context of discovery. That is, scientists might choose their research topic because of the source of funding available, or because they hope that their results will serve some nonscientific goals, or because they had a dream. The method of science, however, is supposed to winnow out these values and biases. In the context of justification, the only thing that should matter is whether the hypotheses generated in the context of discovery are useful empirical hypotheses that can explain and predict the world as it really is.

Social constructivists and feminist philosophers of science reject this distinction. They argue that values and biases are ineliminable from science–even in the process of justification. Thus, they argue that value judgments involved in the acceptance and rejection of theories and hypotheses are "beyond rational criticism," because all rational criticism is from a standpoint, and particular standpoints are defined by some method. When we want to try to support or falsify some hypothesis, we adopt a method of data collection and interpretation. However, any choice of method also stands in need of empirical testing. This cannot be done, they argue, without making further methodological value judgments.

The assumption that the conditions for experiments done in the laboratory are sufficiently similar to conditions in the real world so that judgments about the real world can be based on the experiments is one example of a methodological value judgment; that results of experiments with rats can be extrapolated to humans is another. It is worth quoting some other examples in detail because I intend to focus my critique on this idea of methodological value judgments. Shrader-Frechette writes: "[Risk] Assessors must make value judgments about which data to collect; how to simplify myriad facts into a workable model; how to extrapolate because of unknowns; how to choose statistical tests to be used; how to select sample size; ... [they must] decide where the burden of proof goes, which power function to use, what size of test to run, and which exposure-response model to employ."[58]

We could multiply these examples with ones more specific to models of global systems. For instance, the choice of parameterizations and boundary conditions, and decisions regarding the omission or inclusion of certain feedback processes, would be considered value-laden methodological judgments by Shrader-Frechette and McCoy.

It is often claimed that all theories are value-laden to some extent. When suggesting new avenues of research and ability to make novel predictions, scientists seem to choose to pursue one line of research over another by considering such qualities as simplicity, fit with currently accepted theory, unifying power, and fruitfulness. These preferences express the value scientists place on such things as coherence with their present commitments and so on. Thus, scientists judge that it is good to prefer simple theories over complex ones or that it is right to choose to pursue a line of research on the ground that it promises to generate novel sorts of predictions.[59] It might be thought, then, that the claim of value-ladenness might not be especially devastating to the ecological theories that Shrader-Frechette is discussing, or to models of global systems that are the subject of this book. The existence of value judgments does not necessarily mean that the theories or models chosen are so infected with subjectivity that they are incapable of making objective predictions. Is there something special about ecological theories that makes this objection more telling in their case? Shrader-Frechette would say that there is. The evidence for most ecological theories is uncertain to an extent far greater than usual, and thus the value judgments of ecological theories are much more extensive. This is more than a mere preference for simplicity. Rather, value judgments define the very nature of some theories. Shrader-Frechette and McCoy agree with Nelson Goodman that the prevalence of concepts such as "balance of nature" and "stability" in ecology might be explained by the desire to construct an ecology that can be easily adopted as the basis for various ethical judgments about the environment.[60] Different ethical commitments will be translated into different methodological value judgments, resulting in a different theory about the way the environment operates.

Ignoring the role of methodological value judgments, it is claimed, leads to a false sense of objectivity. The early "Limits to Growth" model has been criticized for ignoring the fact that it is value-laden. Cole writes that it ignored "the real world of social, political and cultural values while making implicit assumptions about these values."[61] Specifically, the model did not include in its simulations any political or economic reactions to future population growth or resource depletion because it was assumed that they would be ineffective in dealing with the predicted crises. This assumption, it is claimed, was a value judgment and not a statement of fact.[62] The "Limits" model, like any model, is structured by many methodological value judgments, some of which are explicit – that there are physical limits, that signals and responses are delayed, and that the system is inelastic when overstressed – and some of which are not explicit – that we should be aiming at a more equitable and just distribution of wealth and that the present balance of nature should be maintained.[63] Almost every methodological value judgment of the model may be questioned. For example, the law of diminishing returns is assumed in many places throughout the model: agricultural yields will not keep increasing proportionally to investment, resources will get more and more expensive to retrieve, pollution will get increasingly more difficult to control. However, all three of these assumptions have been disputed.[64] For example: "One of its [the Limits model's] main modes of collapse is resource depletion. The main reason for this is the assumption of fixed economically-available resources and of diminishing returns in resource technology. Neither of these assumptions is historically valid."[65]

Even the most central assumption of the Limits model has been seriously questioned. U. Columbo has argued against the idea of finite resources: "It is technology which produces the energy – not the raw materials. Therefore, it is possible now to abandon the view of the Earth as a treasure chest, filled with unique resources which, once consumed, can never be replenished."[66] Similarly, the assumption that only capital and manufacturing sectors generate growth has been challenged. Education and research might be more important.[67] Minor changes in any of these assumptions can cause large changes in the model's behaviour. A decision to place more emphasis on, say, the pollution sector of the model, more than another modeller might, will change the nature and time of the predicted crisis.

The above arguments are meant to show that risks can only be assessed against a background of values. It is this necessary inclusion of value judgments in the formation of theories and models that eventually leads Shrader-Frechette and McCoy to conclude that a new rationality must be used for their assessment.

Procedural rationalities are not uncommon. As J.G. March describes them: "Ideas of process rationality emphasize the extent to which decisions find their sense in attributes of the decision process, rather than in attributes of decision outcomes ... Explicit outcomes are viewed as secondary and decision-making becomes sensible through the intelligence of the way it is orchestrated."[68]

Shrader-Frechette and McCoy point out that this idea of an imperfect decision-making procedure matches well with the normal practices of the average reasonable person. When reasonable people are asked to decide whether to sacrifice immediate benefits or risk potential disaster for future generations, they do not make such decisions using ideas of ideal rationality. Rather, they decide on the basis of what they consider it to be rational to fear, on the voluntariness of the risk, its familiarity, its fairness, its "dread factor," the strength of faith they have in experts, and many other nonscientific criteria.[69] Rather than rationality being defined in terms of logic and abstract reason, it is defined by the way people actually make decisions.

Shrader-Frechette and McCoy wish to make it clear that the decisions made using an ethical and political procedure about the acceptability of a risk will sometimes differ from those made solely on the basis of standard accounts of rationality. One such difference reveals itself in what is called the dilemma of standardization. Some risk assessors assume that we should evaluate all kinds of risks with the same measure. For example, if two activities are associated with the same probability of death then we should be equally concerned about both activities. (This is also referred to as the "commensurability presupposition."[70]) On this assumption, the fact that we now attach a great deal more importance to airplane safety than to auto safety is a mistake because the risk of death by auto accident is so much greater than the risk of death by airplane accident. Similarly, if to direct money to low-profile areas of disease prevention can save more lives than spending it on cancer research, then we ought to do so because the value of saving a life should be standard across all risk decisions. If the goal of risk assessment is to save lives, then the specific nature of the risks should not matter. The problem, of course, is that it does matter. The public expresses a demand for disproportionately high safety precautions for nuclear power plants and airplane travel (disproportionate with probability estimates), because probability is not the only factor in risk assessment. Concern for equality, for example, will reinforce what at first might seem a counterintuitive position. Shrader-Frechette argues that while it might seem that standardization treats everyone equally, equality is not equivalent to same treatment – some people are more sensitive to pollution, have special needs, and therefore it does not make sense to treat risks as if they could be easily quantified and weighed by simple addition.[71] The acceptance or rejection of risks will include moral as well as epistemic considerations. The new rationality would take into account the public's fear and distrust of scientific assurances of safety and hence would favour more cautious decisions than the standardization assumption would allow.

Shrader-Frechette recognizes that her suggestion is radical: "To claim that science is laden with value judgments ... is to challenge the whole edifice of epistemology in the West ... We need to build a new account of rationality and objectivity."[72] However, she too does not wish to appear epistemologically relativistic. Objectivity, she claims, is defined in terms of unbiased and open criticism as well as empirical verifiability. She quotes Israel Scheffler to clarify her claim: "Scientific objectivity requires simply the possibility of intelligible debate over the merits of rival paradigms."[73]

Like Seller, she claims that objective knowledge should not be equated with the infallibly true. Rather, objective knowledge is the ability to predict and control events.[74] This kind of knowledge is not relativistic because values *alone* do not determine the facts. Beliefs are open to change in response to new observations.

This account solves three problems. First, it recognizes the need for obtaining a common-denominator value set for making decisions about social risks. Individual risk-taking can make sense because it is based on one coherent self, complete with a fairly consistent set of values. While decisions about individual risks are possible because of the existence of a single, consistent value set, no such set is available to justify public risks. A climber, for instance, can make a life-risking decision to climb Mount Everest because the climber's history and values make such a climb worth the risk. Ian Hacking writes that, as an individual decision-maker, he can adjust probabilities and utilities until they "feel good to me."[75] However, the diversity of values in a community makes decisions about community risk problematic. There is no similar consistent value set in the community that would provide justification for acting in ways that risk global catastrophe, even if the risk were small and the benefits great. In order to supply a coherent value set for public risk-taking, Hacking suggests that we construct a machinery for consensus weighing and comparing of subjective probabilities and attempt a theory of interpersonal utility. A procedural, democratic account of rationality might provide something of the sort. Shrader-Frechette suggests drawing up several risk assessments using various ethical biases and then subjecting these assessments to open, democratic debate.

Second, this account of rationality does reply to suggestions that risk assessment is political. Gross and Levitt and others have criticized supporters of doomsday forecasts for being nonscientific. A procedural, democratic account of rationality explains why this occurs: all science is value-laden.

Third, it comments on the controversy between lay and expert risk assessments. Critics of precaution argue that the layperson's fear of catastrophe is irrational. The average person's psychological make-up tends toward risk-aversion: "In repeated experiments, people show themselves willing to take a considerable gamble to avoid a loss, but are less willing to take the *same* gamble to reap an analogous gain."[76] However, the fact that most people prefer to forgo possible benefits to avoid risking a loss of what they already have is a mere psychological quirk that should be resisted; this is not a rational basis on which to decide general policy, the critics argue. The psychological attitude of most people toward risk is seemingly inconsistent – e.g., we fearlessly hurtle down the highway, but are worried about ozone depletion. We fear unfamiliar risks like HIV more than we should, statistically speaking, and don't fear familiar risks as much as we should. Perhaps we should ignore these psychological attitudes rather than reinforcing them through public policy. However, on the democratic, procedural account of rationality, experts cannot make claim to value-free judgment, and thus their judgments should not be considered more worthy or more accurate than lay judgments that might be influenced by different values.

A CRITIQUE OF DEMOCRATIC EPISTEMOLOGY AND PROCEDURAL RATIONALITY

Environmentalists have been attacked for being irrational and letting nonepistemic values decide what they are to believe about the possibility of environmental disaster. In their defence, Seller, Shrader-Frechette, and McCoy redefine rationality. They want an epistemology where it turns out to be rational to form beliefs in accordance with political or ethical values. However, I think it is a mistake in this case for them to take up the same language in the defence as is used in the attack. Both the attack and the defence confuse epistemology of belief with decisions and risk assessment. I think that both sides of this debate have failed to distinguish between belief and action. Both sides have confused the question of what is the case (What risks attend present trends? What are the probabilities of various outcomes? What are the chances of this course of action resulting in that state of affairs?) with the question of what is to be done – ethically, prudentially, and politically (Should we change our present ways? Should we take this or that course of action?). Shrader-Frechette adopts the language of epistemology, I think, because she sees her view as analogous to an epistemological debate in philosophy of science. I explained on page 127 that Shrader-Frechette bases her position on an analogy she draws between philosophers of science and risk assessors. However, the disanalogy is more telling in this case. Questions about scientific theory choice are epistemological, but questions about models of global risk inevitably involve politics as well as epistemology. Of course, Shrader-Frechette knows this, and it is really the thrust of her whole book. However, I think she inappropriately retains epistemological language in discussions about how to act.

The truth is not affected by what we want to do. Epistemic judgments are not the same as action guiding values. Once you start confusing determinations of what is the case with determinations of what to do, truth becomes relative to a show of hands. It seems wrong to me to let political considerations have any part in decisions about what is the case.

These authors want to explain how scientists with different ethical commitments come up with different policy advice. Scientists concerned with efficiency of conservation dollars give different advice than those whose goal is to preserve the greatest number of species. They argue that the different policies are generated from different epistemic beliefs, that are in turn generated from different methodological value judgments. These different methodological value judgments are the cause of varying interpretations of the same observations. I cannot understand, however, why many of the so-called methodological value judgments are evaluative assumptions instead of factual ones. Why are they methodological judgments instead of methodological principles? Most of the examples given, some quoted above, reveal that alleged methodological value judgments are often nothing but factual assumptions and empirical hypotheses that can be evaluated and tested. The methodological judgments are empirical ones, like the judgment about what constitutes an adequate sample size, and as such they might be correct or incorrect. They are based on experience of past experiments. We know, for example, that political polls with representative samples of about 250 responses are accurate, say, 90 percent of the time. A smaller sample size is less accurate, while a larger sample size does not substantially change accuracy. That judgments of fact were made by whomever made them because of what they valued methodologically is irrelevant to whether the world is as the judgments of fact say it is. They are all factual, empirical claims. They stand or fall with the ability of the theories in which they are found to predict and/or control the course of events. Now, perhaps some sociologist of science will find that this or that particular assumption or hypothesis is employed by a set of scientists who share personal, social, or whatever values, but not by those who don't share these values. Would this mean that values are supplying a bridge between hypotheses and evidence or filling in the gaps of underdetermined models and theories? No. The factors that enable these scientists to cover uncertain territory are empirical assumptions and hypotheses. The dispute between those who employ some assumption and those who reject it is not a dispute over values, but a dispute over what sort of theory or model has the most chance of predictive success: one that includes this assumption or one that does not. That one's values can be cited in an explanation of why one makes the assumptions one does in no way makes one's assumptions anything less than assumptions about the empirical nature of the world. Certainly, ecologists can have a normative program; but that doesn't mean that their descriptions of the world are permeated by their values, at least not in any way that must detract from the scientific nature of their descriptions. Maybe a model was constructed in the hope that we would come to know how a more equitable distribution of wealth could be achieved, or to see that such a distribution would have effects congenial to those who do not value, in itself, the equitable distribution of wealth. The model is at the service of a normative vision, but it is not a slave to this vision. It purports to describe features of the world, features that exist whether anyone likes them or not. Values supply a reason for creating some model; people want to know what is to be done if their normative vision of the world is to be realized. If the notions of "balance of nature" and "stability" are found in theories that enable us to predict and/or control the course of events, so much the better for these notions; if they are found in theories that fail to enable us to do this, so much the worse for them. This is true of any notion seeking accreditation. The source of these notions is entirely irrelevant to the question of their scientific status. Even if one accepts that we cannot always keep science and values separate, we should strive to do so.

Suppose that – in Seller's example of leukaemia, cited earlier – we know that the frequency of leukaemia among malnourished children who live near nuclear power facilities is greater than it is among malnourished children who do not live near nuclear power facilities. Now, among those people concerned about childhood leukaemia, some want more money and effort to go to combatting malnutrition than to combatting nuclear power, others want the opposite. Each can have good and bad reasons for favouring one course of action over the other, and these reasons are composed of *belief and value pairs*. No-one in this example is coming anywhere close to allowing their political commitments to decide their view of the *truth*; people are asking only *what to do*. There is no need to believe that radiation is causing leukaemia, or that it is not. One need only take one of these claims as a working hypothesis. There is no need to construct a new epistemology, but only a need to keep straight which decisions are epistemological and which are political. What a political body can do is to approve or reject a plan that is based on a scientific appraisal of the situation and the chances of success of a course of action – and, of course, it can approve or reject it on the basis of whether it thinks the science any good. How people might vote on the matter, or what they might as a group decide to do, given their accurate or flawed understanding of the risk, has nothing to do with what is the case.

The satisficing account of decision-making described in chapter 7 captures the spirit of Seller, Shrader-Frechette, and McCoy's defence of ethical influences on decisions about uncertain risk without the need for revising the traditional conception of rationality. Just as a criminal trial serves to further goals in addition to finding the correct answer to the question of guilt or innocence (e.g., it serves to show that justice is being done, it provides a deterrent), our prior political commitments will affect our choice of response to global doomsday models because we are trying to further other goals besides the goal of getting the science right.

·10·

A Pascalian Argument for Precaution

The basic thrust of arguments in favour of acting precautiously is that there is too much at stake for us to do otherwise. The disposition to err on the side of caution seems so commonsensical to so many people, that it is often thought not to be in need of justification. When there is a possibility of catastrophe, it is best to eliminate the chances of that possibility. However, this often-stated argument is, perhaps, too vague and simplistic–for reasons outlined in chapter 8. It is rejected by many as unscientific for its reversal of the burden of proof. The principle appears to be unworkable in practice because there are just too many things to be precautious about. Just how seriously are we to take the precautionary principle? Is only the slightest hint of catastrophe sufficient to rule out some activity? Much stronger and more specific arguments, that take into account the difficulties of such blanket caution, are required to justify the enormous costs involved. Some of these arguments will now be examined.

The precautionary principle is not in as bad a shape as we left it in chapter 8. While *epistemological arguments* provide the support for epistemic and practical conservatism, *ethical arguments* often carry the weight for precautionary policies. Despite the problems involved with adopting a precautionary stance, there are many who defend it with moral arguments that, they claim, override the difficulties outlined in chapter 8. Opportunity costs are irrelevant, they argue, because some risks are not morally justifiable. There are ethical reasons for rejecting the accepted practice of theoretical and practical conservatism, for reversing the burden of proof, and instead adopting the precautionary strategy when faced with potential threats. Only by adopting the precautionary principle, it is argued, will we honour deep and important ethical principles such as fairness or respect for autonomy.

Ethical arguments are thought to be effective in reversing the burden of proof. As I argued in chapter 8, the burden of proof is usually shouldered by those who make positive assertions. By insisting that such claims be proven before acceptance, this standard of epistemic conservatism reduces the error of accepting false claims of effect. This practice will always lay the burden of proof on those who are warning of catastrophe, and we have seen how difficult it is to carry this burden. However, there are cases when the burden of proof is carried by those who wish to go against an ethical norm. Simon Blackburn argues that if "in some situation there is a proper presumption that something is true, anyone seeking to prove its opposite is said to bear the burden of proof."[1] Thus, argues James Cargile, we may assert that torture is evil without assuming the burden of proving this. There are many such assertions for which we do not demand proofs. What ethical arguments for the precautionary principle do is provide the context that leads to the "proper presumption" that an attitude of valuing global safety above all else does not acquire a heavy burden of proof. The ethical arguments shift the burden of proof to those who claim that no harmful effects will occur.

Immediately below, I articulate and try to formulate explicitly a few of the ethical arguments that might justify the precautionary principle. Honouring ethical principles in our actions can be costly from the point of view of what we most like or desire to do. For the most part, ethical principles constitute constraints to place on the pursuit of our interests. If one accepts the principle that ethical reasons can trump arguments against precaution, then the arguments about to be given in this chapter will convince. However, ethical arguments can be ineffective in two ways. The first way is within the domain of ethical arguments themselves.

Not all people who take ethics seriously agree on which ethical principles are sound ethical principles. Some will argue that the survival of the human race is an ethical value, even though this conflicts in some cases with taking the survival of individuals as an ethical value. Is it morally justifiable to lessen the quality of life for some individuals, or compromise their security, in the name of protecting the *species* of humanity? Is not "species" merely an abstract concept? Similarly, an ethical principle that inveighs against imposing risks on people against their will might conflict with either of the first two suggested ethical principles. The efficacy of ethical arguments, therefore, will extend only as far as the principle involved meets with acceptance.

The second way in which ethical arguments are inefficacious is that not everybody holds that ethical reasons against doing something override other sorts of reasons for doing it. However, there are prudential reasons to convince those not convinced by ethical arguments for precaution. While ethical principles propose to us what we should want or value, whether we already do want or value it, prudential reasons tell us how best to secure what we already now in fact do want and value. Prudential arguments will appeal to and guide the behaviour of those who value only their own comfort and ability to realize their goals in life. I will conclude by giving prudential arguments that parallel the arguments of Pascal and William James and claim that we cannot afford not to be precautious.

ETHICAL ARGUMENTS FOR PRECAUTION

Preserving the Human Race is an Ethical Duty

Jonas, for one, supplies an ethical argument for precaution by claiming that it is wrong to risk the survival of the human race. In *The Imperative of Responsibility* he argues that it is ethically wrong to put the existence or essence of humanity at risk even when the risk is slight and the gains likely to be got from doing so are large. This ethical principle provides a foundation for precaution despite the lack of factual certainty or reliable scientific models because we can be certain of the moral principle even if we are uncertain about future events. We should not allow uncertainties in predictions of catastrophe to justify inaction because we have a moral duty to give priority to forecasts of fear over those of hope.

Jonas's basic ethical principle is that the existence or essence of humanity is an ultimate good more important than other goods. "Never must the existence or the essence of humanity as a whole be made a stake in the hazards of action," he writes.[2] What Jonas means by the existence of humanity is clear, but what he means by its essence might not be. We are not to stake in the hazards of action the ability of humans to be decent to each other, to love, and to live in freedom. That is, not only must we not, then, risk the existence of humanity, but we must ensure that that existence continues to be dignified. We must not risk the possibility that future persons will live an unacceptable existence because of pollution that we caused, or because we caused the death of most ocean life. This principle implies, continues Jonas, that prognoses of catastrophe should take precedence over prognoses of good fortune. He writes that we should grant "priority to well-grounded possibilities of disaster (different from mere fearful fantasies) over hopes even if no less well-grounded."[3]

Because we must preserve the human race as an ultimate good, we can distinguish legitimate from illegitimate risks of technological activity. We should never take risks simply "because of the enticement of a wonderful future but only under the threat of a terrible future."[4] We can use technology to pursue a good future when doing so does not risk the existence or essence of humanity. But we can also use technology to avoid a humanless or inhumane future even when we thereby risk creating a humanless future ourselves. If the survival or essence of humanity itself is at stake, then it is legitimate to act in ways that might themselves risk the existence of humanity. He suggests that this might sometimes be the case in warfare. If there is a "threat of a terrible future," it might be ethically acceptable to risk the survival of humanity because the terrible future is unacceptable.

Still, it would be unethical to risk the ultimate good for progress, for comfort, for efficiency, for an improvement in our condition. It would be unethical to risk the survival of humanity even if the goal were utopian, and we thought the utopian goal is within reach. It is wrong to try to reach for utopia if we could fail and failure would mean disaster. We could survive without the utopia, without the added progress, he points out, so it is not necessary, and hence unethical, to take the risk. We can live without taking risks to gain further benefits, and we should not jeopardize anything that is not surplus.[5]

Many critics of global models of catastrophe point out that the models predicting catastrophe must inevitably fail to include scientific breakthroughs of the type that have always saved us in the past – breakthroughs in efficiency, or new energy sources, for example. Jonas's argument implies, however, that it is not right to count on such things, for we would be worse off if our future plans included such bonuses and yet they failed to materialize. He gives a principled reason for ignoring the possibility that whatever problem we notice will receive an as yet unknown technological fix. Thus, even if we recognize the possibility of adapting to, say, global warming by building sea-walls and launching artificial satellite-clouds, it would be ethically wrong to adopt that strategy.

Jonas's defence of precautionary thinking was groundbreaking. However, it is unlikely to be very effective at convincing others of the wisdom of the precautionary principle. There are, in the main, two reason for this. First, it is not always clear how the principle of priority should be applied and, second, his claim that the preservation of the human race is the ultimate good is defended with controversial reasoning.

Jonas argues that we have a moral obligation to give priority to "well grounded fears" over hopes. This does not really solve one of the main problems with the precautionary principle, however, because it still leaves the question of what constitutes "well-groundedness." If what we mean by "well grounded" is "supported with scientific evidence," then there really is no need for a precautionary principle. The precautionary principle was supposed to provide reasons for action even in the absence of scientific reasons.

It is still unclear just which risks are justified and which are not. For example, genetically modifying our food supply might be creating a risk of catastrophe, some say. Exactly what the risks are supposed to be remains a little vague, but it is not merely fear of the unknown that lies behind the debate over genetically modified food. Rather, the fear is based on analogies with other experiences. Scientists did not foresee any health problems with cloning animals, for example. DNA is DNA, it is commonly argued. Yet unexplained health problems have recently surfaced in many cloned animals. Similar, unexpected health problems might arise from ingesting genetically modified foods. We are conducting an unethical experiment on the human race, some worry.

Now, does Jonas give us any guidance here? He argues that we should not risk the existence or essence of humanity merely to ameliorate our circumstances. Thus, it might be argued that genetically modifying food is unjustified because its purpose is merely to improve production efficiencies and profit. However, he notes that his principle allows us to take risks in order to avoid an unacceptable future–and this is precisely the reason given by some for justifying the continuation of genetic engineering of foods. Wouldn't the invention of a rice variety that is highly nutritious, grows in poor soil, and requires little pesticide use be an advancement toward the elimination of the risk of an unacceptable future of undernutrition or starvation for much of the world's burgeoning population? Jonas assumes that all technological risks are undertaken for progress rather than to "alleviate what is unbearable," and hence it follows that great risks of technology should not be undertaken.[6] It would be easy to take issue with his assumption, though.

Jonas's defence of his ultimate principle regarding the ethical priority of preserving the human race is unusual. In attempting to provide a metaphysical grounding for this ethical principle, he gives reasons that probably do not have widespread support. He claims, for example, that the existence of humanity implies that we have a moral duty to preserve the existence of moral agents.[7] Justifying this principle as deeper than any other principle such that it overrides all others will be difficult. Another ethical principle, that according to which we are to maximize expected utility over all affected by our actions, seems also to be deep and ultimate, and yet it might well counsel us to stake humanity's existence if the payoff of our action is large and chance of failure is small.

He argues that the disappearance of humanity would be an infinite loss,[8] but critics would wish to know for whom it would be a loss. A generally accepted reason to avoid catastrophe would be to avoid the pain and suffering that it would bring. But once the catastrophe has eliminated all of us, there remains no point of view from which to give the loss an evil aspect. He claims that humanity has "no right to suicide,"[9] but his reasons for this are unclear. He makes an analogy with the caring of children. He argues that parents do not require reasons and proof to take precautionary action with their children. Similarly, we should recognize the sanctity of life and act precautiously to preserve it, without the need for proof.

Jonas's reasoning in favour of the precautionary principle will have little appeal, even among those who think a cogent ethical defence of the precautionary principle is possible. Jonas's defence is a defence that raises difficult and controversial issues in the ontology of value, issues on which it is unreasonable to expect any consensus soon to emerge. One might well on reflection find his arguments persuasive, but one might just as well not. I myself might tend to agree with him that there is value in preserving the existence of humanity, but would most others? This defence of the precautionary principle will only convince the like minded.

First, Do No Harm

In *Method in Ecology*, K.S. Shrader-Frechette and E.D. McCoy elaborate an ethical argument for precaution to the conclusion that it is wrong to risk catastrophic harm even when the risk is slight and the benefits to be obtained are great. They claim that the duty of nonmaleficence, the duty not to cause harm, is stronger than the duty of beneficence, the duty to prevent or remove harms, or to promote the good. If this is so, then the precautionary principle is justified. We have a duty to stop our harmful activities, to prevent harms, and no appeal to improved welfare can override this duty. It is much more important to protect what we have than to risk it for a chance of improving our general welfare.

They also note that most of us accept this claim that the right to be protected from harm is greater than the right to promote a positive good. "Most political theorists," they write, "regardless of their persuasion, would probably agree that it is prima-facie more important to protect the public from serious harm (e.g., loss of species, nuclear accidents) than to enhance welfare (e.g., by permitting land development, by providing electrical power on demand). This is in part because the right to protection against those who cause harm is more basic than rights to welfare. It is more important, in society, to prevent acts that cause serious harm than to promote acts that enhance welfare."[10] When someone has a right, someone has a duty to respect that right. Since this claim about the relative strength of the two duties – the duty not to cause harm and the duty to promote the good, is widely accepted – so too should be the precautionary principle.

We can explain why we have these ethical intuitions psychologically by noting that, generally speaking, we prefer to keep what we have to risking it on something new, but this does not vindicate these attitudes. In any case, since this argument begins from a common ethical intuition, the argument to the precautionary principle has the practical virtue of resting on a widely accepted first principle, a virtue that Jonas's argument, above, lacks.

There are probably many clear cases of actions that promote welfare while risking the production of harm. In these cases, people will have little difficulty understanding which specific course of action is, according to the principle, the ethically sound course of action. People who accept the principle, as most of us do, will have a clear sense of what to do. However, it is not always easy to tell, and sometimes not even possible, whether an action counts as forestalling harm rather than as promoting welfare, and vice versa. A distinction is often made between negative duties, i.e., not causing harm, and positive duties, i.e., promoting welfare, but in practice it is difficult to apply this distinction. Many activities that are associated with a risk of disaster can be described as activities meant to forestall harm, so the duty to privilege the avoidance of harm is unclear. Genetic engineering of crops, for example, represents a possible harm. It is possible that the introduction of new genetic arrangements into the ecosystem will upset that ecosystem in unforeseen and undesirable ways. Yet, the genetic engineering of crops to increase resistance to pests and to increase yields might be described as forestalling the harm that would result from food shortages. People's willingness to take these risks derives not from the ethically unacceptable desire to gamble with harms to improve their condition, but from the ethically respectable concern to avoid harms.

Furthermore, sometimes the attempt to avoid one harm is precisely what creates another harm. It is therefore very difficult to decide just which actions are justified by the principle of avoiding harm. The principle could not provide guidance, for example, on the energy policy decision between the risk of global warming through the use of fossil fuels or the risk of nuclear accidents and poisoning from the use of nuclear energy. The ethical theory of utilitarianism copes with this problem by claiming that the right action to perform is the one most likely to maximize benefit. The only ethically sound thing to do in any situation is simply to maximize expected utility. This principle of prioritizing nonmaleficence cannot make use of this probabilistic strategy, though. It requires that we not risk harm despite the odds. Furthermore, the risks we have been considering are uncertain–it is precisely because we do not know the probabilities involved that makes it a unique problem.

The principle also runs into problems with the distinction between social harms and individual harms. Inoculating children, for example, reduces and sometimes eliminates the risk that they will contract a serious illness. But the act of inoculating a child itself puts the child at risk of harm. Given that most of the children in a community are inoculated, the risk that the next child will be harmed by the disease if not inoculated might be lower than the risk posed by the inoculation. Socially, the duty seems to be to innoculate each individual child in order to reduce harm in the children as a whole. Individually, however, it seems that the duty is to avoid the inoculation in order to do no harm to the individual child. Consider parents who properly estimate that the risk of harm to their child from inoculation approaches the risk of harm from contracting the disease and who then, acting on their duty not to expose a child to harm, refuse to allow their child to be inoculated. One's intuition here might be that the parents' action is ethically wrong, for in refusing to allow their child to be inoculated they are being unfair to other parents or children and perhaps exploiting them for selfish benefits.

Public Risks Are Unfair

Individual risk-taking is often a highly valued behaviour. We admire those who take chances and poke fun at those who are overly cautious. Those who don't take risks are considered boring or trapped in a shell that limits their opportunities. A safe life can be pleasant, but mediocre. Risk-taking can be fun in itself, and even when the personal risk borders on the irrational and reckless, such as climbing Mount Everest or skiing across Antarctica, we nonetheless find this behaviour acceptable and even admirable. Apart from thrill-seeking, risk-taking can also sometimes be a rational part of pursuing one's ends. Career choices, opting for one medical treatment over another, and many other of life's decisions involve the taking of considered risks. We accept both types of risk-taking as part of everyday life and value it as an important factor in personal growth.

Public risk-taking is a different matter. While no-one would think to justify putting the public at risk for the thrill the public might get, it might appear perfectly reasonable for a society to take risks as part of its pursuit of social goods. However, it is argued by some that the fact that public risk would be spread widely over many people somehow makes it unacceptable to take it. D.C. Mayo and D. Hollander write "When the issue involves the lives, health and well-being of so many people, it seems extremely difficult to justify any other strategy [than exercising maximum caution]."[11]

It is not obvious, though, that public risks are unacceptable just because the interests, health, and well-being of many people are involved. If the social goal is respectable and the decision to pursue it through risking something was made democratically, then this seems consistent with other values we hold. Democratic decisions are made, for example, about the fluoridation of public drinking water or levels of chlorine. These decisions involve risks to the health of entire communities. Such risk-taking would only be legitimate, however, when the costs and risks are borne fairly. An ethical principle of fairness would imply that those who bear the risk of some endeavour ought to receive the benefits of that endeavour. And those who reap the benefits of success would be those who pay the costs or assume the risk of bad consequences. This is not usually the case for the behaviours that involve global risks. They are not such that the costs and risks could often be distributed fairly, but would almost inevitably fall also on people who do not benefit from the behaviour. Since this is the case for almost all environmental risks, argue many, including Shrader-Frechette and McCoy[12] and Gordon and Suzuki,[13] we should minimize these risks. Thus, we have another ethical reason for adopting the precautionary principle.

It is reasonable to believe that the distribution of risks and benefits is unfair for most, if not all, of the catastrophic risks considered in this book. Think, for example, of ozone layer depletion. The risks of skin cancer, and other risks associated with the increase in UV radiation, apply to every inhabitant on the planet, in the present and in the near-future. However, the benefits accrued from the use of CFCs for refrigeration, air-conditioning, and cleaning of silicon chips for electronic equipment manufacture, have accrued mainly to people in the developed world.

Industrial pollution is distributed globally, imposing risks on those who have received no benefits from the industry. Gordon and Suzuki describe, for example, how the Canadian Arctic is polluted despite the lack of industry: "Industrial effluents but no industry; PCBs but no electrical transformers; DDT but no crop pests; cesium but no nuclear generators."[14]

The risks from sea-level rise associated with global warming are similarly distributed unequally. The industries and activities responsible for producing greenhouse gases benefit mainly those nations that are not seriously at risk from rising sea-levels. Rich, industrialized countries could more readily adapt to changes in sea-level by the construction of dikes and sea-walls in sensitive areas. This might not be a practical solution in all cases; however, it is not even conceivable in the poorer countries that are most at risk. In Bangladesh, millions of people live at sea-level who are already vulnerable to storms and floods that take thousands of lives. According to Gordon and Suzuki, "Europe, America, and what used to be called the Soviet bloc countries generate 71 percent of the world's carbon dioxide. Sea-level rise and drought in the next five decades could drive more than 60 million people from their homes in countries such as Bangladesh, Egypt, the Maldives."[15]

Acceptance of the principle of fairness together with these facts, though, will not justify that risks ought to be minimized. It seems only to imply that the benefits of risky endeavours should go to those who were put at risk. The principle of fairness could be used in arguments in favour of distributing industrial wealth in ways that are more fair to developing nations. It does not seem to supply the justification needed for the precautionary principle, though taking it seriously would likely mean fewer risks levelled on fewer people.

Public Risks Are Involuntary

Even a fairly distributed risk could well be a risk it would be immoral to take, however, if the costs and risks fall on people who did not decide to take the risk. An ethical reason for adopting the precautionary principle is that it is immoral to impose risks on those who do not consent to them. Individuals who assume risks do so voluntarily, and the risks are limited to those individuals. Public risks might not be voluntarily assumed, but, rather, imposed on people whether or not they have consented. It is the lack of consent that concerns us about the examples described in the previous section. The inhabitants of the Arctic did not consent to the health risks imposed on them by global industrial pollution. Sea-level dwellers did not consent to the risk of sea-level rise from global warming. Even if they are compensated, therefore, in an effort to distribute the benefits of the risk fairly, it would not satisfy the ethical requirement that risks should not be imposed without the consent of those at risk.

Whether someone has consented to be put at risk, however, might be open to challenge. By willingly accepting the benefits of an activity people are implicitly consenting to be put at risk and, hence, can legitimately be put at risk. So, one can accept the principle that it is unjust to impose risks on people who have not consented, but deny that the people put at risk have not accepted to be at risk.

This argument for implied consent has a long history and diverse application. However, it is controversial, especially in the cases we have been considering, that people are willing to accept benefits in exchange for taking risks. They might accept the benefits only because it is an offer they cannot refuse. Such bargains are not fair deals. This line of argument has to do with the conditions under which the decision to pay the costs and take the risks is made. If the decision was properly made in accordance with sound democratic procedures, and if the community recognizes a right of exit so that those who disapprove of the laws or policies democratically enacted can remove themselves from their jurisdiction, and the costs and risks are distributed fairly, then all those who do not remove themselves from the community do implicitly consent to be placed at risk. This ideas seems right in most cases of communal or corporate endeavour. If you dislike a properly made decision, you can sell your shares or leave the group. If you do not sell your shares or leave, you have implicitly consented.

The analogy with most communal and corporate endeavours fails in the cases of global risk. Countries are not relevantly like clubs or companies because the right of exit has no application in a world of sovereign states and poverty-stricken populations. Because this acceptance of risks in exchange for benefits is an offer that cannot be refused, the argument for implied consent is not adequate.

A further problem that this defence for the precautionary principle shares with the previous ones, has to do with deciding priority rankings for various other relevant ethical principles. The ethical principle defended in this section, that it is wrong to put at risk people who do not freely consent to be put at risk, can conflict with other equally sound ethical principles in application to cases. It is wrong to put people at risk without their consent, but it is also good to fly, and drive, and create energy for many other activities. It might be next to impossible to perform any activity without putting at risk some who have not consented. In an imperfect and imperfectable world, the best that can be done is to compensate those wrongly harmed when put at risk without their consent.

As has been noted, however, in cases of catastrophic global risk the concepts of insurance and compensation for loss are inapplicable. Would it not be wrong then, to stake in a risk something that cannot be compensated for if lost? In the next section, I will argue that the relevant point here is whether we do so *knowingly.*

Culpable Ignorance

I argued earlier that ignorance alone was insufficient to justify the precautionary principle because this would count against all activity whatsoever. Every action we take might have a bad result. However, a certain kind of ignorance might justify precaution– culpable ignorance.

Annette Baier thinks that our moral decisions are structured around "what maddens us" rather than around the most effective way of preserving lives.[16] She claims, for example, that we focus on deviant and malicious behaviour rather than on pollution even though pollution is probably the cause of more deaths. Negligence and culpability "madden" us in this sense and, perhaps, provide the strongest ethical reason for precaution. The basic notion of moral culpability is that, if a harm occurs, a harm that should have been foreseen and could have been avoided, then a basic moral obligation has been violated. One is ethically culpable for the bad effects of one's actions just so long as one could have known that they posed a risk. If this principle holds, then those whose activities bring on global catastrophe will be ethically culpable for them, for they should right now know that their activities are risky, even if they cannot know how risky.

K.R. Blair and W.A. Ross have argued that our continued production of greenhouse gases is a case of morally culpable negligence. Since suffering would result if the models of greenhouse gas induced global warming are accurate, then "all action that can be taken to avoid global warming is ethically required."[17] René Dumont blames the greenhouse effect for droughts causing one million deaths from starvation each year.[18] He then goes on to accuse suspected greenhouse offenders of murder. P. Danielson argues that, while we may be excused for greenhouse gas emissions to date, we are "no longer innocent" of the potential effects of our actions and are morally responsible for any harm that might result.[19] Danielson thinks that the awareness of a plausible model that predicts harm is sufficient to make us morally responsible. In one sense he is right. If we pursue activities, knowing that they might result in harm to others, we are morally culpable. However, Danielson and others might be begging the question with respect to global warming. Whether the emission of greenhouse gases is a risky activity is the very issue in question. Can people ignorant of the complicated, long-term effects of their actions really be held morally responsible? It seems wrong, for example, to blame the inventor of CFCs for his lack of foresight about the connection between CFCs and ozone depletion. The benefits of CFC use were obvious and immediate, while the risks were unknown, complex, and distant.

We need some criterion of culpable ignorance that prevents both excessive caution and total inaction when faced with ignorance. While not arguing explicitly for the precautionary principle, Ian Hacking uses the notion of "culpable ignorance" as a measure to determine when we are morally required to be wary of the risks of high technology.[20] He argues that acting in a state of ignorance is morally culpable only if harm is likely to result and the agent *could have* found out about the likely consequences of the action. However, an agent is not morally culpable if no reasonable person could be expected to know these consequences. We cannot be expected to know all the harms to which some of our current actions might be linked. Obviously, many of our actions will have negative consequences of which we remain ignorant, so we will cause harm. However, only some of this ignorance is morally culpable.

Adapting arguments from Aristotle and Aquinas, Hacking argues that culpability involves universals rather than particulars.[21] There are universal generalizations about car brakes, for instance, that one should be expected to know. However, one cannot be expected to know the particular details of any one set of brakes that might result in harm. Thus, I am morally blameworthy if I drive a car without getting the brakes checked regularly, but not if some unusual mechanical defect affects my particular car. We should be expected to know in general what kinds of effects will result from familiar types of actions, even if we can't predict the exact details, but ignorance of the future is too widespread to be allowed as a general reason for inactivity. Thus, we should not limit our activities except in cases where our actions would be culpable. To think otherwise and to behave with maximum caution, constantly worried about what disaster might occur, is to assume, unreasonably, an unlimited responsibility for all world events resulting from human activity.

However, there is a type of situation in which an agent is blameworthy for acting under ignorance even when the ignorance is unavoidable. Hacking argues that the history of large-scale technology provides enough examples to give rise to a universal belief in the possibility of harm resulting from new and unfamiliar technology. It is the record of historical failures that justifies precautionary action and nullifies the excuse of ignorance in any particular case.

Interference effects, for example, are unpredictable in detail, but are a familiar enough phenomenon that we would be culpable not to expect them. Interference effects are those that result from the combination of technologies; for example, while we might know how two medicines work individually, we do not know what interference effects will result when these medicines act in concert. Hacking supposes that it might be "in the nature of rapidly expanding deployment of new technologies to produce such [harmful] effects by unforeseeable universal facts."[22] There is enough regularity to such harm that it should now be expected. A universal expectation would provide grounds for precaution even though it is impossible to find out whether any particular new technology is likely to result in harm. Hacking claims that his argument provides "a new kind of reason for being leery of new technology."[23]

The precautionary principle is founded on similar expectations of harm resulting from actions that have global effects – expectations formed from historical experience. What Hacking's reasoning implies is those who perform actions that have potentially catastrophic consequences can be morally culpable even if we cannot foresee the specific consequences. They are culpable because experience has shown that we should expect harmful consequences from certain kinds of activity. We know from experience that lack of precaution will cause the continued degradation of the environment. PCBs, CFCs, and DDT were all assumed to be harmless. We know now that scientists were wrong in all three cases. Hacking's argument for culpability takes into account the pervasiveness of these errors. Although we might not be able to justify precaution in any one instance, precaution is justified as a general strategy because it is based on our previous experiences. By accepting an overall strategy of epistemic and practical conservatism, we are culpable because this makes it more likely that we will eventually make an error and fail to take precautionary action when it is necessary.

Jonas, too, uses the notion of culpable ignorance to argue for precaution. In normal situations, we have a duty to find out what are the likely effects of our actions. However, he writes, "a novel moral problem" is created when our ability to predict the consequences of our actions is weak, and the potential effects of those actions are global, irreversible, and cumulative. In unprecedented situations, he continues, the "recognition of ignorance becomes the obverse of the duty to know."[24]

Despite the distinction between ignorance in general and culpable ignorance, the principle against acting if risk is foreseeable, Hacking concedes, might still lead to "inaction, or at least severely curtailed endeavor." He admits that this conclusion is "plainly wrong" – meaning, I think, that excessive caution is not practical.[25] However, he suggests no way out of this dilemma except to say that neither cost-benefit analyses nor ethics will suffice.

PRUDENTIAL REASONS FOR PRECAUTION

The criticisms I have raised against the ethical arguments for precaution are of three types. First, there are different reasons for holding that the precautionary principle is an ethically sound principle, and sometimes these reasons are incompatible. The duty to preserve humanity, for example, is not necessarily compatible with the duty to distribute risks and benefits fairly or with the duty not to impose risks on those who do not consent to them. Second, I have outlined criticisms that have to do with applying the precautionary principle to courses of action and that uncover difficulties in getting firm direction from the precautionary principle in individual cases. Perhaps this criticism has been too broad. It might be that there are no ethical principles that can always be applied to cases unambiguously and then acted on without creating further difficulties. Perhaps this counts against all ethical principles. This leads to the third criticism, which is that there are those who simply do not agree that ethical considerations should outweigh other considerations.

If one is already in the realm of the ethical, the above arguments for precaution will be convincing. However, these arguments for precaution are as controversial as the principle itself. How do we convince those who reject these ethical arguments for precaution, that precaution is still the wisest course of action? I will argue that some kind of prudential reasoning must suffice. Even a fairly distributed risk, assumed democratically, could well be a risk it would be prudentially foolish to take. Some risks are such that they would be stupid to take, even if not immoral.

Shrader-Frechette, too, has argued practical reasons for preferring precaution. One practical reason is that precautionary measures are easier to figure out than how best to measure and balance costs and benefits.[26] Ranking outcomes in order of preference to find out which is the worst is much easier to do than estimating probabilities and assigning values to those outcomes. A distrust in the ability of bureaucracies to actually make correct cost-benefit analyses is another reason she gives to avoid that method.[27]

I will argue in this direction and develop even stronger reasons for precaution from the perspective of prudence. I will borrow arguments from Pascal and William James to argue that taking precautions is the prudential thing to do. Following Pascal, I argue that if one allows the perceived high costs of precaution to sway against precautionary action, then one just does not understand the infinite costs of catastrophe. Following William James, I argue that deciding a course of action on nonintellectual grounds, which the arguments for precaution specifically call for, is justified under special conditions – conditions that define what James calls a genuine option.

I have described the possibility of global catastrophe as a zero-infinity dilemma (very small probability, but immeasurable stakes). Perhaps the most famous zero-infinity dilemma is Pascal's Wager. Pascal argued that it is better to be safe than sorry when it comes to wagering on the existence of God. The probability of God's existence is unknown, but even if there is the smallest probability of God's existence it would be wise to bet on it because the potential rewards are infinite. If God exists, and you have failed to bet correctly, then you will have lost infinite rewards. However, if God does not exist, you will have lost relatively little on this wager.

Pascal's Wager runs smoothly if we assume a low cost to the safe option. He assumed that acting on this wager would cost little. Many disagree. Choosing to live as if God existed will affect every aspect of one's life. Pascal would reply, however, that if you think those costs are high, you do not understand the infinite rewards that might result from a correct wager on God's existence.

Environmental zero-infinity dilemmas are concerned with global catastrophe, while Pascal's argument concerns the loss of infinite rewards. Nonetheless, the principles governing the rationality of gambling in the two cases remain the same. Even if the costs of precautionary action to avoid ozone depletion, global warming, nuclear winter, and overpopulation are high, they are nothing compared to the infinite negative value of a catastrophe. We might have to restructure society to use less energy. We might have to live a lower-technology lifestyle. We might have to spend billions. But all this is nothing compared with catastrophe. In a sentence that applies equally well to both Pascal's dilemma and ours, Pascal admits that, if one follows the advice of his argument, one cannot "enjoy noxious pleasures, glory and good living." However, he suggests that we might be compensated with other kinds of pleasures.[28]

Importantly, James Wernham points out that, despite the usual interpretation of Pascal's argument, it is not about believing, but about gambling.[29] When we make the wager in favour of God's existence we are acting – not believing. One need not believe in God to gamble that God exists. What one risks in making this bet is a way of living, not an epistemic belief.[30] We saw in the previous chapter that Seller, Shrader-Frechette, McCoy, and others have made the mistake of thinking that one must believe in predictions of environmental disaster in order to act on them. However, environmental zero-infinity dilemmas are no different from Pascal's wager when it comes to the absence of the need for belief. There is no need to believe that models of global systems predicting catastrophe are true in order to bet on them. Acting precautiously in response to some model that is predicting global catastrophe is a gamble, not a belief in the truth of that model.

We will see that William James, too, fails to distinguish action from belief. Wernham argues that, while James's famous essay *The Will to Believe* is usually interpreted as an argument about the ethics of belief, and this is the way James himself describes it, it is better reconstructed as an argument for the prudence of acting one way rather than another. Thus, it is not really about either ethics or belief. *The Will to Believe* is often misinterpreted as concerning these things, Wernham says, because James took himself to be responding to arguments of Huxley and Clifford – arguments about the ethics of belief. Clifford's doctrine is that it is ethically wrong to believe anything on less than sufficient evidence, for to believe on less than sufficient evidence is always, though often only indirectly, to risk harming people.

Rehearsing a version of the argument for epistemic conservatism, James comments that scientific reasoning exhorts us never to believe something on insufficient evidence. James quotes Clifford to illustrate this point of view: "If [a] belief has been accepted on insufficient evidence ... it is ... *stolen* ... [Our] duty is to guard ourselves from such beliefs as from a pestilence ... It is wrong always, everywhere, and for every one, to believe anything upon insufficient evidence."[31] Thus, Clifford would argue that it is wrong to believe things merely because of political or other desires to believe things.

Nonetheless, James attempts to show that there are certain momentous decisions that must be made under uncertainty and, in these cases, the caveat of scientific reasoning to suspend judgment is misplaced. James argues that, under certain conditions, it would be foolish not to have faith in something even when science does not supply the required evidence.

Although James admires the virtues of epistemic conservatism and the "thousands of disinterested moral lives ... [of] patience and postponement"[32] dedicated to it, he argues that for some issues we reach a point where suspending judgment is foolish. James writes:

> Believe truth! Shun error! – these, we see, are two materially different laws; and by choosing between them we may end by coloring differently our whole intellectual life. We may regard the chase for truth as paramount, and the avoidance of error as secondary; or we may, on the other hand, treat the avoidance of error as more imperative, and let truth take its chance. Clifford ...

> exhorts us to the latter course. Believe nothing, he tells us, keep your mind in suspense forever, rather than by closing it on insufficient evidence incur the awful risk of believing lies ... For my own part, I have also a horror of being duped; but I can believe that worse things than being duped may happen to a man in this world: so Clifford's exhortation has to my ears a thoroughly fantastic sound. It is like a general informing his soldiers that it is better to keep out of battle forever than to risk a single wound.[33]

James thinks that those who must wait for certainty are suffering from "mental weakness" caused by a belief that "there is something called scientific evidence by waiting upon which they shall escape all danger of shipwreck in regard to truth. But there is really no scientific or other method by which men can steer safely between the opposite dangers of believing too little or believing too much. To face such dangers is apparently our duty, and to hit the right channel between them is the measure of our wisdom."[34]

Usually, James notes, our decision to suspend judgment is of no consequence.

"Wherever the option between losing truth and gaining it is not momentous, we can throw the chance of *gaining truth* away, and at any rate save ourselves from any chance of *believing falsehood*, by not making up our minds at all till objective evidence has come ... in human affairs in general, the need of acting is seldom so urgent that a false belief to act on is better than no belief at all ... seldom is there any such hurry about them that the risks of being duped by believing a premature theory need be faced."[35] However, there are occasions, ones which present what James calls "genuine options," when we cannot afford to wait for definite evidence.[36] A genuine option is a choice between live hypotheses that is forced, momentous, and cannot be decided on intellectual grounds.

A live option is simply one where each of the choices presents itself as a real possibility. For a New Englander of the late nineteenth century, for example, the option of being either a Christian or an agnostic is an option that has some rational appeal. However, the option of being a theosophist or a Mohammedan is not a live option, says James, because it is too unfamiliar to be considered as a real possibility.

An option is forced if it is either unavoidable or if we cannot afford not to decide. Suspension of belief is not, then, an option. If we run the same risk by suspending judgment about belief in God, for example, as we do by deciding against God's existence, then the option is forced. Even if the option is undecidable on intellectual grounds, one is forced to decide anyway – on what James called "passional" grounds.

An option is momentous if the stakes are high, or the decision is unique or irreversible. James supposes that an invitation to join Nansen's now famous expedition to the North Pole would be a momentous option. It is a unique opportunity for a chance at immortality. At the same time, this expedition is life-risking. Once decided, it is irreversible – Nansen's ship, the *Fram*, is going to sail with or without you.

James goes on to propose that moral and religious questions are examples of genuine options that "immediately present themselves as questions whose solution cannot wait for sensible proof."[37] In these situations, contrary arguments might seem equally plausible. James thinks that there are equally good arguments on both sides of the free will/determinism debate, for example. Similarly, the arguments for and against belief in God might be equally strong. Whenever we are in a situation where we are undecided because of equally good arguments on both sides, and yet we must decide – because the decision is forced and momentous – then we must decide on nonintellectual grounds: "pure insight and logic ... are not the only things that really do produce our creeds."[38]

There are many decisions in life that must be made on nonintellectual grounds. Our belief that there is such a thing as truth, and that we can attain it, is, he thinks, a "passionate affirmation of desire." Nancy Cartwright has argued that another such choice is whether to adopt the traditional scientific faith in an orderly universe, one governed by laws that we can discover, rather than adopt a faith in an "untidy," chaotic universe.[39] The choice we make between risking either the loss of truth or the inclusion of falsehood is also a passionate decision, rather than one made on intellectual grounds.

James concludes from all this that, in these types of cases, one has a right to believe, even on insufficient evidence. Wernham, however, questions whether James's arguments really support the right to belief in God rather than simply the prudence of maintaining a working hypothesis about the existence of God and being ready to act on that hypothesis. James's argument makes more sense, says Wernham, if seen as a Pascalian argument about wagering. On Wernham's sympathetic reconstruction, James's argument supports only the claim that it is sometimes prudent to act on less than sufficient evidence. The prudent course is to act, for example, as if God existed, whether one believes or not.

First, Wernham notes that James misrepresents Clifford as advising that we wait "in suspense forever" when there is insufficient evidence; Clifford's argument recommends only suspension of belief, not suspension of action. Clifford never argues that we should discontinue investigations or that some other action would be unjustified.

Second, a forced option is one where we are forced to act, not to believe. One can never be forced to believe. Furthermore, a forced action is not necessarily tied to belief. I might make my wager because I think I have some reason to believe one thing rather than another, but I could make the same wager without believing anything. I might merely entertain the relevant hypothesis. Wernham argues that James fails to distinguish these important concepts, equivocating between belief, faith, gambling, and accepting something as a working hypothesis as a method of obtaining further evidence.[40] On one occasion, James even compares believers to gamblers at a racetrack, making an obvious equivocation between believing and acting.

As already mentioned, Wernham thinks that James adopts this language of ethics and belief to the detriment of his argument, because he gets caught up in responding to Clifford's argument, which is a moral argument about duty and belief. Clifford is arguing that suspending judgment in the face of uncertainty is the morally correct thing to do. Furthermore, belief was central to the genesis of James's argument. Early versions of the Will-to-Believe argument relied heavily on the notion of self-verifying beliefs. In one early-version example, a stranded mountain climber might die unless he believes he can make a difficult leap. James argues that the climber ought to believe he can make the leap even if he has little grounds for this belief because believing so improves the chance that the leap will be successful. Other examples include a lover whose love will never be requited unless he or she has faith that it will be requited. A belief can sometimes contribute to its coming true. The notion of self-verifying beliefs plays only a small role in the final version of the Will-to-Believe argument, however. It is not suggested, for example, that belief in God will contribute to the truth of that belief. Nonetheless, argues Wernham, James retained the notion of belief even when it was no longer useful to his argument.

Pascal and James's arguments provide reasons for action – not belief. They provide reasons for acting on the hypothesis that God exists. Both Pascal and James point out that the benefits of this action will have an immediate pay-off and might even have an infinite pay-off in the afterlife. These are reasons for faith. Wernham points out that faith is not belief: faith is to make a gamble. To have faith in a religious hypotheses is not to believe that that hypothesis is true, but rather to bet. Faith is not a belief, but rather an action, the action of risking something on an option when one does not know whether that option is true and might not even believe that it is. James's argument reveals that there are cases where it would be imprudent not to have faith.

The question of what to do in response to models of global systems that predict catastrophe also fits James's criteria for a genuine option. Models present live options and we are forced to make momentous decisions. The evidence from the models does not provide sufficient evidence for belief, and this is why they fail to convince everyone. However, they do provide reasons for action.

Part I of this book argued that models of global systems represent possibilities but cannot yet supply reliable probabilities. Whether the predictions of global catastrophe are reliable cannot, then, be decided on intellectual grounds. Nonetheless, the models present live options. Even if we admit that global models are tautologies, we can still use them as indicators of logical possibilities. So long as models and predictions are plausible, they are live options.

Decisions about how to act in the face of predictions of global catastrophe are forced because they require us to act immediately, or at least very soon, and we cannot escape them because inaction is itself a decision. If the models are correct, the longer we wait, the more difficult it will be to take action to prevent catastrophe.

Decisions involving global risk might also be momentous. In some cases, precautionary action does not require much of a gamble. If cheap and available alternatives to CFCs or fossil fuels are available, then we ought to use them. However, when such cheap options are not available, the precautionary wager will be much higher. Regardless of the cost, however, the wager is momentous because, it is argued, the future of the planet is at stake. The consequences are potentially irreversible. Even if the probability of some catastrophe is low, any wager involving that possibility will be momentous.

In the special case of a genuine option, James claims that it is foolish to adopt the ways of the epistemological and practical conservative, because these ways were designed for cases of ordinary decision-making. When evidence alone is unable to supply answers, one *must* act for other reasons. So, a proponent of the precautionary principle might argue that there are good reasons for acting precautiously, and for not acting in ways that might be risking the future of life on Earth –

reasons not based on estimates of the reliability of predictions. We must choose to act one way or the other in reaction to claims about the need for precautionary action based on models of global systems, but that does not mean we are forced to believe or not believe in these models. We are still free to suspend belief and adopt the semantic view of models described in chapter 2 – namely, merely accept the models as working hypotheses.

Notes

PREFACE

1 Emerson, *Nature*, 230.
2 Pascal, *Pensées*, 153.

INTRODUCTION

1 Gordon and Suzuki, *It's a Matter of Survival*, 3.
2 Swift, *Gulliver's Travels*, 206.
3 *The Economist*, vol. 328, no. 7828, 11 September 1993: 13.
4 Chang, "Cassini Safely Passes Earth," 2.

CHAPTER ONE

1 Thompson and Schneider, "Nuclear Winter Reappraised," 984.
2 Turco et al., "Nuclear Winter: Global Consequences of Multiple Nuclear Weapons Explosions," 1283–92.
3 Thompson and Schneider, "Nuclear Winter Reappraised," 984–5.
4 Rothman, "A Memoir of Nuclear Winter," 111 and 126.
5 Rothman, "A Memoir of Nuclear Winter," 130.
6 Rothman, "A Memoir of Nuclear Winter," 128.
7 Thompson and Schneider, "Nuclear Winter Reappraised," 991–2.
8 Thompson and Schneider, "Nuclear Winter Reappraised," 984–5.
9 Thompson and Schneider, "Nuclear Winter Reappraised," 987.
10 Thompson and Schneider, "Nuclear Winter Reappraised," 993.
11 Thompson and Schneider, "Nuclear Winter Reappraised," 983.
12 Meadows et al., *Beyond the Limits*, 141–60.

13 Meadows et al., *Beyond the Limits*, 152–3.
14 Meadows et al., *Beyond the Limits*, 150.
15 Meadows et al., *Beyond the Limits*, 148.
16 Meadows et al., *Beyond the Limits*, 150.
17 Meadows et al., *Beyond the Limits*, 152.
18 Meadows et al., *Beyond the Limits*, fig. 5.5, 152.
19 Simon, *Population Matters*; Simon and Kahn, *The Resourceful Earth*; North, *Life On a Modern Planet*.
20 Sjoberg, *Risk and Society*, 75.
21 Meadows et al., *Beyond the Limits*, fig. 4.8.
22 Meadows et al., *Beyond the Limits*, xiii.
23 Meadows et al., *Beyond the Limits*, xiv.
24 Meadows et al., *Beyond the Limits*, 148.
25 Meadows et al., *Beyond the Limits*, 179.
26 IPCC, "Climate Change 2001: The Scientific Basis," 13.
27 IPCC, "Climate Change 2001: The Scientific Basis," 12–16.
28 Balling, *The Heated Debate*; Michaels, *Sound & Fury*.
29 Houghton et al., *Climate Change: The IPCC Scientific Assessment*, xii.
30 Buckmaster, "The Arctic – A Canadian Case Study," 67.
31 Balling, *The Heated Debate*, 35.
32 Bernard, *Global Warming Unchecked*, 12.
33 Meadows et al., *Beyond the Limits*, 92; Bernard, *Global Warming Unchecked*, fig. 1.1, 11.
34 Bernard, *Global Warming Unchecked*, fig. 1.1.
35 Bernard, *Global Warming Unchecked*, 5, quoting from the *Bull. of American Meteorological Society*.
36 Balling, *The Heated Debate*, xxvi.
37 Buckmaster, "The Arctic – A Canadian Case Study," 78.
38 Bernard, *Global Warming Unchecked*, 58.
39 Buckmaster, "The Arctic – A Canadian Case Study," 67.
40 Bernard, *Global Warming Unchecked*, 82.
41 Buckmaster, "The Arctic – A Canadian Case Study," 73.
42 Bernard, *Global Warming Unchecked*, 92.
43 Buckmaster, "The Arctic – A Canadian Case Study," 63.
44 Balling, *The Heated Debate*, 45.
45 Houghton et al., *Climate Change: The IPCC Scientific Assessment*, xxvii.
46 Balling, *The Heated Debate*, xxiii.
47 Charlson et al., "Climate Forcing by Anthropogenic Aerosols," 423–30.
48 Matthews, "The Rise and Rise of Global Warming," 6.
49 Isaksen, "Dual Effects of Ozone Reduction," 322–3.
50 Houghton et al., *Climate Change: The IPCC Scientific Assessment*, xviii.

51 Schneider, "The Changing Climate," 75.
52 Houghton et al., *Climate Change: The IPCC Scientific Assessment*, xx.
53 IPCC, "Climate Change 2001: The Scientific Basis," 9.
54 Shackley et al., "Uncertainty, Complexity and Concepts of Good Science in Climate Change Modelling: Are GCMs the Best Tools?" 8.
55 Hare, "The Challenge," 20.
56 Hare, "The Challenge," 19; see also Schneider, "The Rising Seas," 112–17.
57 McKibben, *The End of Nature*, 14.
58 Houghton et al., *Climate Change: The IPCC Scientific Assessment*, xvii.
59 McKibben, *The End of Nature*, 14.

CHAPTER TWO

1 Hesse, *Models and Analogies in Science*, 4.
2 Bernard, *Global Warming Unchecked*.
3 Rothman, "A Memoir of Nuclear Winter," 114.
4 Hesse, *Models and Analogies in Science*, chapter 1; also Achinstein, *Concepts of Science*, chapter 7.
5 Popper, *The Open Universe*, 44.
6 Levins, "The Strategy of Model Building in Population Biology," 421–1.
7 Peters, *A Critique For Ecology*, 106–10.
8 Shackley et al., "Uncertainty, Complexity and Concepts of Good Science in Climate Change Modelling: Are GCMs the Best Tools?" 163–70.
9 Shackley et al., "Uncertainty, Complexity and Concepts of Good Science in Climate Change Modelling: Are GCMs the Best Tools?," 169.
10 Cartwright, *How the Laws of Physics Lie*, 4.
11 Cartwright, *How the Laws of Physics Lie*, 143–62.
12 Cartwright, *How the Laws of Physics Lie*, 145.
13 Cartwright, *How the Laws of Physics Lie*, 128–42.
14 Cartwright, *How the Laws of Physics Lie*, 144 and 158.
15 Giere, *Explaining Science*, 82.
16 Giere, *Explaining Science*, 47.
17 Giere, *Explaining Science*, 48.
18 Giere, *Explaining Science*, 80–2.

CHAPTER THREE

1 Giere, *Understanding Scientific Reasoning*: second edition, 142.
2 Giere, *Understanding Scientific Reasoning*: second edition, 118.
3 Balling, *The Heated Debate: Greenhouse Predictions versus Climate Reality*, 100.

CHAPTER FOUR

1 Gates et al., "Validation of Climate Models," 100.
2 Houghton et al., *Climate Change: The IPCC Scientific Assessment,* xxxviii-xxxix.
3 Balling, *The Heated Debate,* 55.
4 Balling, *The Heated Debate*, 55–65.
5 IPCC, "Climate Change 2001: The Scientific Basis," 2.
6 Oster, "Predicting Populations," 831–44.
7 Peters, *A Critique For Ecology,* 125.
8 Shackley et al., "Uncertainty, Complexity and Concepts of Good Science in Climate Change Modelling: Are GCMs the Best Tools?" 170.
9 Shackley et al., "Uncertainty, Complexity and Concepts of Good Science in Climate Change Modelling: Are GCMs the Best Tools?" 168.
10 Houghton et al., *Climate Change: The IPCC Scientific Assessment*, xxviii.
11 Gates et al., "Validation of Climate Models," 97n1.
12 Houghton et al., *Climate Change: The IPCC Scientific Assessment*, 9.
13 Gleick, *Chaos*, 44.
14 Houghton et al., *Climate Change: The IPCC Scientific Assessment*, xviii.
15 Cartwright, *How the Laws of Physics Lie*, 11.
16 Cartwright, *How the Laws of Physics Lie*, 12.
17 Hacking, "Culpable Ignorance of Interference Effects," 144.
18 Hacking, "Culpable Ignorance of Interference Effects," 145.
19 Houghton et al., *Climate Change: The IPCC Scientific Assessment*, xxiii.
20 Mesarovic, *Mankind at the Turning Point*: first edition, 37.
21 Mesarovic, *Mankind at the Turning Point*: first edition, 39.
22 McCutcheon, *Limits to a Modern World*, 82–3.
23 Cole et al., *Thinking About the Future*, 10.
24 Thompson and Schneider, "Nuclear Winter Reappraised," 988.
25 Balling, *The Heated Debate*, xxiv and 103.
26 Epstein, "Is Global Warming Harmful to Health?" 50–7.
27 Barthes, *Empire of Signs*, 15–18.

CHAPTER FIVE

1 Bernard, *Global Warming Unchecked*, 85.
2 Gates et al., "Validation of Climate Models," 101.
3 Gates et al., "Validation of Climate Models," 102–20.
4 Matthews, "The Rise and Rise of Global Warming," 6.
5 Houghton et al., *Climate Change: The IPCC Scientific Assessment*, xxviii.
6 Roberts, *Modelling Large Systems*, 8.
7 Cole et al., *Thinking About the Future*, 113–14.

8 Hughes, *World Modeling*, 184.
9 Giere, *Understanding Scientific Reasoning*: second edition, 142.
10 Shackley et al., "Uncertainty, Complexity and Concepts of Good Science in Climate Change Modelling: Are GCMs the Best Tools?" 170 and 174.
11 Balling, *The Heated Debate*, 32.
12 Bernard, *Global Warming Unchecked*, 12.
13 Bernard, *Global Warming Unchecked*, 12.
14 Bernard, *Global Warming Unchecked*, 88.
15 McKibben, *The End of Nature*, 10.
16 Meadows et al., *Limits to Growth*, 22.
17 McCutcheon, *Limits to a Modern World*, 101.
18 Bernard, *Global Warming Unchecked*, 79.
19 Rothman, "A Memoir of Nuclear Winter," 124–32.
20 Rothman, "A Memoir of Nuclear Winter," 124.
21 Cole, *Models of Doom*, 133.
22 Balling, *The Heated Debate*, 33.
23 Shackley et al., "Uncertainty, Complexity and Concepts of Good Science in Climate Change Modelling: Are GCMs the Best Tools?" 184.
24 Bernard, *Global Warming Unchecked*, 103.
25 Houghton et al., *Climate Change: The IPCC Scientific Assessment*, xxv-xxvi.
26 Rothman, "A Memoir of Nuclear Winter," 115.
27 Houghton et al., *Climate Change: The IPCC Scientific Assessment*, fig. 2, xv.
28 Cubasch and Cess, "Processes and Modelling," 83.
29 Bernard, *Global Warming Unchecked*, 83.
30 Schneider, "The Changing Climate," 75.
31 Schneider, "The Changing Climate," 75.
32 Schneider, "The Changing Climate," 75–6.
33 Schneider, "The Changing Climate," 76.
34 Schneider, "The Changing Climate," 76.
35 Watson et al., "Greenhouse Gases and Aerosols," 17.
36 Balling, *The Heated Debate*, 47–9.
37 Balling, *The Heated Debate*, 69–70.
38 IPCC, "Climate Change 2001: The Scientific Basis," 10–11.
39 Balling, *The Heated Debate*, 97–119.
40 Balling, *The Heated Debate*, 95.

CHAPTER SIX

1 Peters, *A Critique For Ecology*, 38–73.
2 Shrader-Frechette and McCoy, *Method in Ecology*, 6.

3 Quine, "Two Dogmas of Empiricism," 20–46.
4 Lakatos, "Falsification and the Methodology of Scientific Research Programmes," 182.
5 Cartwright, *How the Laws of Physics Lie*, 56.
6 Cartwright, *How the Laws of Physics Lie*, 3.
7 Cartwright, *How the Laws of Physics Lie*, 13.
8 Peters, *A Critique For Ecology*, 196.
9 Cartwright, *How the Laws of Physics Lie*, 111.
10 Thompson, *The Structure of Biological Theories.*
11 Thompson, *The Structure of Biological Theories*, 104.
12 Bunge, *Method, Model, and Matter*, 31.
13 Bunge, *Method, Model, and Matter*, 38.
14 Bunge, *Method, Model, and Matter*, 41.
15 Shackley et al., "Uncertainty, Complexity and Concepts of Good Science in Climate Change Modelling: Are GCMs the Best Tools?" 162–80.
16 Peters, *A Critique For Ecology*, 147–77.
17 Peters, *A Critique For Ecology*, 36.
18 Peters, *A Critique For Ecology*, 191–3.
19 Peters, *A Critique For Ecology*, 60.
20 Peters, *A Critique For Ecology*, 172–4.
21 *The New Yorker*, 16 December 1996: 71.

CHAPTER SEVEN

1 Hurka, "Ethical Principles," 23–39.
2 Mayo and Hollander, *Acceptable Evidence*, 189.
3 Rescher, *Risk*, 97–8.
4 Giere, *Understanding Scientific Reasoning*: second edition, 145.
5 Worzel, R. "The Sins of Eco-phonies," A26.
6 Elster, *Nuts and Bolts for the Social Sciences*, 160.
7 Giere, *Explaining Science*, 7.
8 Giere, *Explaining Science*, 161.
9 Giere, *Explaining Science*, 21.

CHAPTER EIGHT

1 Giere, *Understanding Scientific Reasoning*: second edition, 350.
2 *The Economist*, "The Hard Rain," 81.
3 von Furstenberg, *Acting Under Uncertainty*, preface; Shrader-Frechette, *Risk and Rationality*, 131–45; Dixon and Massey Jr, *Introduction to*

Statistical Analysis, fourth edition.

4 Gould, "The Piltdown Conspiracy," 201–40.

5 Giere, *Understanding Scientific Reasoning*: second edition, 105.

6 Shrader-Frechette and McCoy, *Method in Ecology*, 157.

7 Cargile, "On the Burden of Proof," 61.

8 Buckmaster, "The Arctic–A Canadian Case Study," 76.

9 Bernard, *Global Warming Unchecked*, 157.

10 Stebbing, "Environmental Capacity and the Precautionary Principle," 292.

11 Thoreau, *Walden*, 114.

12 Rescher, *Risk*, 75–6.

13 Rescher, *Risk*, 94.

14 Shrader-Frechette, *Risk and Rationality*, 188–90.

15 Shrader-Frechette, *Risk and Rationality*, 112–16.

16 Sagoff, "The Limits of Cost-Benefit Analysis," 76.

17 Hacking, "Culpable Ignorance of Interference Effects," 153.

18 Hacking, "Culpable Ignorance of Interference Effects," 143.

19 Hacking, "Culpable Ignorance of Interference Effects," 152.

20 Hill, "Regulating Biotechnology," 177–8.

21 Plato, *Apology*, 29a-b, 15.

22 McKibben, *The End of Nature*, 68–9; Peacock, "The Ozone Surprise," 23.

23 Vanier, *In Weakness, Strength*, 38–9.

24 Bankes, "International Responsibility," 115–17.

25 Earll, "Commonsense and the Precautionary Principle – An Environmentalist's Perspective," 184.

26 Rescher, *Risk*, 73.

27 Danielson, "Personal Responsibility."

28 Miles Jr, *Awakening from the American Dream*, 191.

29 VanderZwaag, *CEPA and the Precautionary Principle Approach*, 11.

30 Bankes, "International Responsibility," 122.

31 VanderZwaag, *CEPA and the Precautionary Principle Approach*, 4.

32 Milne, "The Perils of Green Pessimism," 35.

33 Earll, "Commonsense and the Precautionary Principle – An Environmentalist's Perspective," 183.

34 VanderZwaag, *CEPA and the Precautionary Principle Approach*, 15.

35 VanderZwaag, *CEPA and the Precautionary Principle Approach*, 11.

36 VanderZwaag, *CEPA and the Precautionary Principle Approach*, 4.

37 O'Riordan and Jordan, "The Precautionary Principle in Contemporary Environmental Politics," 193.

38 North, *Life On a Modern Planet*, 257.

39 North, *Life On a Modern Planet*, 257.

40 Hill, "Regulating Biotechnology," 177–8.
41 Earll, "Commonsense and the Precautionary Principle – An Environmentalist's Perspective," 184.
42 Danielson, "Personal Responsibility," 82.
43 Michaels, *Sound & Fury*, 6.
44 Houghton et al., *Climate Change: The IPCC Scientific Assessment*, xi.
45 *The Economist*, "Cool Costing," 84.
46 Steger, quoted in Michaels, *Sound & Fury*, xiii.
47 van Kooten, "Effective Economic Mechanisms: Efficiency and Ethical Considerations," 138.
48 Weyant, "The Costs of Carbon Emissions Reductions," 193.
49 van Kooten, "Effective Economic Mechanisms: Efficiency and Ethical Considerations," 138.
50 Schneider, "The Changing Climate," 79.
51 O'Riordan and Cameron, eds, *Interpreting the Precautionary Principle*.
52 Michaels, *Sound & Fury*, 7.
53 Weyant, "The Costs of Carbon Emissions Reductions," 193.
54 Blair and Ross, "Energy Efficiency at Home and Abroad," 156.
55 van Kooten, "Effective Economic Mechanisms: Efficiency and Ethical Considerations," 133–48.
56 Blair and Ross, "Energy Efficiency at Home and Abroad," 150.
57 *The Economist*, "Cool Costing," 84.
58 Danielson, "Personal Responsibility," 90.
59 Stebbing, "Environmental Capacity and the Precautionary Principle," 289.
60 Peterman and M'Gonigle, "Statistical Power Analysis and the Precautionary Principle," 231–4.
61 Stebbing, "Environmental Capacity and the Precautionary Principle," 288.
62 Stebbing, "Environmental Capacity and the Precautionary Principle," 288 and 291.
63 Cargile, "On the Burden of Proof," 59.
64 IPCC, "Climate Change 2001: The Scientific Basis," 12.
65 Bankes, "International Responsibility," 125.
66 Bankes, "International Responsibility," 123.
67 Michaels, *Sound & Fury*, xiv.
68 North, *Life on a Modern Planet*, 256.
69 Milne, "The Perils of Green Pessimism," 36; Lawrence and Taylor, "Letters," 598–9.
70 Johnston and Simmonds, "Letters," 402.
71 Milne, "The Perils of Green Pessimism," 37.

72 Gray, "Statistics and the Precautionary Principle," 175.
73 Rescher, quoted in Shrader-Frechette, *Risk Analysis and Scientific Method,* 128.
74 Rescher, *Risk*, 114–15.
75 Milne, "The Perils of Green Pessimism," 34–7.
76 Milne, "The Perils of Green Pessimism," 36–7.
77 Gray, "Statistics and the Precautionary Principle," 174.
78 Gray,"Statistics and the Precautionary Principle," 174.
79 Gray, "Statistics and the Precautionary Principle," 176.
80 Gray, "Letters," 599.

CHAPTER NINE

1 Wernham, *James's Will-to-Believe Doctrine: A Heretical View*, 75–80.
2 Longino, "Can There Be A Feminist Science?" 260.
3 Rothman, "A Memoir of Nuclear Winter," 109–47.
4 Freeman Dyson, quoted in Rothman, "A Memoir of Nuclear Winter," 136.
5 Meadows et al., *Beyond the Limits*, 141–60.
6 Meadows et al., *Beyond the Limits*, 152–3.
7 Meadows et al., *Beyond the Limits*, chapter 5.
8 See, especially, Leslie, *The End of the World.*
9 Boehmer-Christiansen, "A Scientific Agenda for Climate Policy?" 400–2; Boehmer-Christiansen, "Global Climate Protection Policy: The Limits of Scientific Advice, part 1" and "Global Climate Protection Policy: The Limits of Scientific Advice, part 2."
10 Boehmer-Christiansen, "Global Climate Protection Policy: The Limits of Scientific Advice, part 2," 192.
11 Boehmer-Christiansen, "A Scientific Agenda for Climate Policy?" 401.
12 Boehmer-Christiansen, "A Scientific Agenda for Climate Policy?" 402.
13 Boehmer-Christiansen, "Global Climate Protection Policy: The Limits of Scientific Advice, part 2," 185.
14 Boehmer-Christiansen, "Global Climate Protection Policy: The Limits of Scientific Advice, part 1," 140–59.
15 Michaels, "Crisis in Politics of Climate Change Looms on Horizon," 14–23.
16 Wildavsky, quoted in Balling, *The Heated Debate,* xxxi.
17 Cole et al., eds, *Thinking About the Future*, 178.
18 O'Riordan and Jordan, "The Precautionary Principle in Contemporary Environmental Politics," 191–3.

19 Jonas, *The Imperative of Responsibility.*
20 Joni Mitchell, "Big Yellow Taxi," in *Ladies of the Canyon*, Reprise Records, 1970.
21 Thompson and Schneider, *Nuclear Winter Reappraised*, *Foreign Affairs*, 1005.
22 Jonas, *The Imperative of Responsibility*, 27.
23 Jonas, *The Imperative of Responsibility*, 27.
24 Stephen Schneider, quoted in Schell, "Our Fragile Earth," 47.
25 Gross and Levitt, *Higher Superstition*, 166.
26 Shrader-Frechette and McCoy, *Method in Ecology*, 85.
27 Rothman, "A Memoir of Nuclear Winter," 137.
28 Gross and Levitt, *Higher Superstition*, 156.
29 Gross and Levitt, *Higher Superstition*, 159.
30 Boehmer-Christiansen, "Global Climate Protection Policy: The Limits of Scientific Advice, part 2," 200.
31 Boehmer-Christensen, "Global Climate Protection Policy: The Limits of Scientific Advice, part 2," 196–9.
32 Kierkegaard, *Either/Or*, 103.
33 Sagoff, "Fact and Value in Environmental Science," 99–116.
34 Gray, "Statistics and the Precautionary Principle," 176.
35 Lawrence and Taylor, "Letters," 599.
36 Earll, "Commonsense and the Precautionary Principle – An Environmentalist's Perspective," 182.
37 Shrader-Frechette, *Risk and Rationality*, 7–8.
38 Shrader-Frechette, *Risk and Rationality*, 68.
39 Seller, "Realism versus Relativism," 173–4.
40 Douglas, Book Review of *Risk and Rationality*, 485.
41 Seller, "Realism versus Relativism," 178.
42 Seller, "Realism versus Relativism," 170–1.
43 Seller, "Realism versus Relativism," 173–4.
44 Seller, "Realism versus Relativism," 174.
45 Seller, "Realism versus Relativism," 169.
46 Seller, "Realism versus Relativism," 169.
47 Seller, "Realism versus Relativism," 169.
48 Seller, "Realism versus Relativism," 173.
49 Seller, "Realism versus Relativism," 175–6.
50 Shrader-Frechette and McCoy, *Method in Ecology*, 141.
51 Shrader-Frechette and McCoy, *Method in Ecology*, 191.
52 Shrader-Frechette, *Risk and Rationality*, 77; Shrader-Frechette and McCoy, *Method in Ecology*, 105.

53 Shrader-Frechette, *Risk and Rationality*, 192.
54 Shrader-Frechette and McCoy, *Method in Ecology*, 88 and 109.
55 Shrader-Frechette and McCoy, *Method in Ecology*, 83–5.
56 Shrader-Frechette and McCoy, *Method in Ecology*, 94.
57 Shrader-Frechette, *Risk and Rationality*, 43.
58 Shrader-Frechette, *Risk and Rationality*, 57.
59 Quine and Ullian, *The Web of Belief*: second edition.
60 Shrader-Frechette and McCoy, *Method in Ecology*, 48.
61 Cole, *Models of Doom*, 202.
62 McCutcheon, *Limits to a Modern World*, 99.
63 Meadows et al., *Beyond the Limits*, 139.
64 Cole, *Models of Doom*, 56; McCutcheon, *Limits to a Modern World*, 71 and 96
65 Cole, *Models of Doom*, 41.
66 Pestel, *Beyond the Limits to Growth*, 120 (quoting U. Columbo).
67 Cole, *Models of Doom*, 74.
68 March, "Bounded Rationality," 148.
69 Shrader-Frechette, *Risk and Rationality*, 49–50.
70 Shrader-Frechette, *Risk and Scientific Method*, 55.
71 Shrader-Frechette, *Risk and Rationality*, 60.
72 Shrader-Frechette, *Risk and Rationality*, 54.
73 Shrader-Frechette, *Risk and Rationality*, 52.
74 Shrader-Frechette, *Risk and Rationality*, 195.
75 Hacking, "Culpable Ignorance of Interference Effects," 141.
76 Mills, "Why Life is Disappointing," 564.

CHAPTER TEN

1 Blackburn, *The Oxford Dictionary of Philosophy*, 51.
2 Jonas, *The Imperative of Responsibility*, 37.
3 Jonas, *The Imperative of Responsibility*, 32.
4 Jonas, *The Imperative of Responsibility*, 36.
5 Jonas, *The Imperative of Responsibility*, 121.
6 Jonas, *The Imperative of Responsibility*, 36.
7 Jonas, *The Imperative of Responsibility*, 10.
8 Jonas, *The Imperative of Responsibility*, 33.
9 Jonas, *The Imperative of Responsibility*, 36–7.
10 Shrader-Frechette and McCoy, *Method in Ecology*, 159–60.
11 Mayo and Hollander, *Acceptable Evidence*, 190.
12 Shrader-Frechette and McCoy, *Method in Ecology*, 162.

13 Gordon and Suzuki, *It's a Matter of Survival*, 133–49.
14 Gordon and Suzuki, *It's a Matter of Survival*, 70.
15 Gordon and Suzuki, *It's a Matter of Survival*, 134.
16 Baier, "Poisoning the Wells," 49–74.
17 Coward, *Ethics & Climate Change*, 9.
18 René dumont, quoted in Picard, "Greenhouse Effect Blamed for Deaths of One Million in Third World Last Year," 35.
19 Danielson, "Personal Responsibility," 87–8.
20 Hacking, "Culpable Ignorance of Interference Effects," 153.
21 Hacking, "Culpable Ignorance of Interference Effects," 137–9.
22 Hacking, "Culpable Ignorance of Interference Effects," 143.
23 Hacking, "Culpable Ignorance of Interference Effects," 152.
24 Jonas, *The Imperative of Responsibility*, 8.
25 Hacking, "Culpable Ignorance of Interference Effects," 153.
26 Shrader-Frechette, *Risk and Rationality*, 126.
27 Shrader-Frechette, *Risk and Rationality*, 126.
28 Pascal, *Pensées*, 153.
29 Wernham, *James's Will-to-Believe Doctrine: A Heretical View*, 75–80.
30 Wernham, *James's Will-to-Believe Doctrine: A Heretical View*, 77.
31 James, "The Will to Believe," 8.
32 James, "The Will to Believe," 7.
33 James, "The Will to Believe," 18–19.
34 James,"The Will to Believe," x-xi.
35 James,"The Will to Believe," 19–20.
36 James, "The Will to Believe," 22.
37 James,"The Will to Believe," 22.
38 James, "The Will to Believe," 11.
39 Cartwright, *How the Laws of Physics Lie*.
40 Wernham, *James's Will-to-Believe Doctrine: A Heretical View*, 101.

Bibliography

Achinstein, P. *Concepts of Science*. Baltimore, Maryland: Johns Hopkins Press, 1968.

Baier, A. "Poisoning the Wells" in *Values at Risk*, D. MacLean, ed. New Jersey: Rowan and Allanheld, 1986, 49–74.

Balling, R.C., Jr. *The Heated Debate: Greenhouse Predictions versus Climate Reality*. San Francisco: Pacific Research Institute for Public Policy, 1992.

Bankes, N. "International Responsibility," in *Ethics & Climate Change*, H. Coward and T. Hurka, eds. Waterloo, Ontario: Wilfrid Laurier University Press, 1993, 115–32.

Bernard, H.W. Jr. *Global Warming Unchecked: Signs to Watch For.* Bloomington and Indianapolis: Indiana University Press, 1993.

Blackburn, S. *The Oxford Dictionary of Philosophy.* Oxford: Oxford University Press, 1994.

Blair, K.R. and W.A. Ross. "Energy Efficiency at Home and Abroad," in *Ethics & Climate Change*, Harold C. and T. Hurka, eds. Waterloo, Ontario: Wilfrid Laurier University Press, 1993, 149–64.

Boehmer-Christiansen, S.A. "A Scientific Agenda for Climate Policy?" in *Nature*, vol. 372, no. 6505, 1 December 1994: 400–2.

Boehmer-Christiansen, S.A. "Global Climate Protection Policy: The Limits of Scientific Advice, part 1," in *Global Environmental Change*, vol. 4, no. 2, 1994: 140–59.

Boehmer-Christiansen, S.A. "Global Climate Protection Policy: The Limits of Scientific Advice, part 2," in *Global Environmental Change*, vol. 4, no. 3, 1994: 185–200.

Buckmaster, H.A. "The Arctic–A Canadian Case Study," in *Ethics & Climate Change*, H. Coward and T. Hurka, eds. Waterloo, Ontario: Wilfrid Laurier University Press, 1993, 61–80.

Bunge, M.A. *Method, Model, and Matter.* Dordrecht: Reidel, 1973.

Cargile, J. "On the Burden of Proof," in *Philosophy*, vol. 72, no. 279, 1997: 59–83.

Cartwright, N. *How the Laws of Physics Lie.* Oxford: Clarendon Press, 1983.

Chang, K. "Cassini Safely Passes Earth," in ABCNEWS.com, 18 August 1999.

Charlson, R.J. et al. "Climate Forcing by Anthropogenic Aerosols," in *Science*, no. 255, 1992: 423–30.

Cole, H.S.D. ed. *Models of Doom.* New York: Universe Books, 1973.

Cole, H.S.D., C. Freeman, M. Jahoda, and K.L.R. Pavitt, eds. *Thinking About the Future.* London: Sussex University Press, 1973.

Conrad, J. *Lord Jim.* New York: Signet Classic, 1961.

Coward, H. and T. Hurka, eds. *Ethics & Climate Change.* Waterloo, Ontario: Wilfrid Laurier University Press, 1993.

Cubasch, U. and R.D. Cess. "Processes and Modelling," in *Climate Change: The IPCC Scientific Assessment*, Houghton et al., eds. Cambridge: Cambridge University Press, 1990, 69–91.

Danielson, P. "Personal Responsibility," in *Ethics & Climate Change*, H. Coward and T. Hurka, eds. Waterloo, Ontario: Wilfrid Laurier University Press, 1993, 81–98.

Dixon, W.J. and F.J. Massey Jr. *Introduction to Statistical Analysis*: fourth edition. New York: McGraw Hill, 1983.

Douglas, M. Book review of *Risk and Rationality: Philosophical Foundations for Populist Reforms* (K.S. Shrader-Frechette, 1991), in *American Political Science Review*, vol. 87, no. 2, June 1993: 485.

Earll, R.C. "Commonsense and the Precautionary Principle – An Environmentalist's Perspective," in *Marine Pollution Bulletin*, vol. 24, no. 4, 1992: 182–6.

The Economist. "Cool Costing," in vol. 326, no. 7801, 6 March 1993: 84.

The Economist. "The Threat from Space," in vol. 328, no. 7828, 11 September 1993: 13.

The Economist. "The Hard Rain," in vol. 328, no. 7828, 11 September 1993: 81.

Elster, J. *Nuts and Bolts for the Social Sciences.* Cambridge: Cambridge University Press, 1989.

Emerson, R.W. "Nature," in *Human Life and the Natural World,* Owen Goldin and Patricia Kilroe, eds. Peterborough, Ontario: Broadview Press, 1997, 225–37.

Epstein, P.R. "Is Global Warming Harmful to Health?" in *Scientific American*, vol. 283, no. 2, August 2000: 50–7.

Gates, W.L., P.R. Rowntree, and Q.C. Zeng. "Validation of Climate Models," in *Climate Change: The IPCC Scientific Assessment*, Houghton et al., eds. Cambridge: Cambridge University Press, 1990, 97–130.

Giere, R.N. *Understanding Scientific Reasoning*: second edition. Toronto: Holt, Rinehart and Winston, 1984.

Giere, R.N. *Explaining Science*. Chicago: University of Chicago Press, 1988.

Gleick, J. *Chaos*. London: Penguin Books, 1987, 44.

Gordon, A. and D. Suzuki. *It's a Matter of Survival*. Toronto: Stoddart, 1990.

Gould, S.J. "The Piltdown Conspiracy," in *Hen's Teeth and Horse's Toes*. New York: W.W. Norton & Company Inc., 1983, 201–40.

Gray, J.S. "Statistics and the Precautionary Principle," in *Marine Pollution Bulletin*, vol. 21, no. 4, 1990: 174–6.

Gray, J.S. Letter in *Marine Pollution Bulletin*, vol. 21, no. 12, December 1990, 599.

Gross, P.R. and N. Levitt. *Higher Superstition: The Academic Left and Its Quarrels with Science*. Baltimore: Johns Hopkins University Press, 1994, chapter 6, 166.

Hacking, I. "Culpable Ignorance of Interference Effects," in *Values at Risk*, Douglas MacLean, ed. New Jersey: Rowman & Allanheld Publishers, 1986, 136–54.

Hare, F.K. "The Challenge," in *Ethics & Climate Change*, H. Coward and T. Hurka, eds. Waterloo, Ontario: Wilfrid Laurier University Press, 1993, 11–22.

Hesse, M. *Models and Analogies in Science*. South Bend, Indiana: University of Notre Dame Press, 1966.

Hill, J. "Regulating Biotechnology," in *Interpreting the Precautionary Principle*, T. O'Riordan and J. Cameron, eds. London: Cameron and May, 1994, 177–8.

Houghton, J.T., G.J. Jenkins, and J.J. Ephraums, eds. *Climate Change: The IPCC Scientific Assessment*. Cambridge: Cambridge University Press, 1990.

Hughes, B. *World Modeling*. Massachuset: Lexington Books, 1980.

Hurka, T. "Ethical Principles.in *Ethics & Climate Change*, H. Coward and T. Hurka, eds. Waterloo, Ontario: Wilfrid Laurier University Press, 1993, 23–39.

IPCC. "Climate Change 2001: The Scientific Basis," Summary for Policymakers: A Report of Working Group I. Approved by IPCC member governments in Shanghai in January 2001.

Isaksen, I.S.A. "Dual Effects of Ozone Reduction," in *Nature*, vol. 372, no. 6504, 24 November 1994: 322–3.

James, W. "The Will to Believe," in *The Will to Believe and Other Essays in Popular Philosophy*. New York: Longmans, Green and Co., 1927, 1–31.

Johnston, P. and Simmonds, M. Letter in *Marine Pollution Bulletin*, vol. 21, no. 8, 1990: 402.

Jonas, H. *The Imperative of Responsibility*. Chicago: University of Chicago Press, 1984.

Kellert, S.H. *In the Wake of Chaos*. Chicago: University of Chicago Press, 1993.

Kierkegaard, S. *Either/Or*, in *A Kierkegaard Anthology*, R. Bretall, ed. Princeton, New Jersey: Princeton University Press, 1973, 19–108.

Lakatos, I. "Falsification and the Methodology of Scientific Research Programmes," in *Scientific Knowledge*, J.A. Kourany, ed. Belmont, California: Wadsworth, 1987.

Lawrence, J. and D. Taylor. Letter in *Marine Pollution Bulletin*, vol. 21, no. 12, December 1990: 598–9.

Leslie, J. *The End of the World: The Science and Ethics of Human Extinction*. London and New York: Routledge, 1996.

Levins, R. "The Strategy of Model Building in Population Biology," in *American Scientist*, vol. 54, no. 4, 421–31.

Longino, H. "Can There Be a Feminist Science?" in *Women, Knowledge and Reality*, A. Garry and M. Pearsall, eds. New York: Routledge, 1996, 251–63.

March, J.G. "Bounded Rationality, Ambiguity, and the Engineering of Choice," in *Rational Choice*, Jon Elster, ed. New York: New York University Press, 1986, 142–70.

Matthews, R. "The Rise and Rise of Global Warming," in *New Scientist*, 26 November 1994: 6.

Mayo, D.C. and D. Hollander, eds. *Acceptable Evidence*. Oxford: Oxford University Press, 1991.

McCutcheon, R. *Limits to a Modern World*. London: Butterworths, 1979.

McKibben, B. *The End of Nature*. New York and Toronto: Anchor Books, 1990.

Meadows, D.H., D.L. Meadows, J. Randers, and W. Behrens. *Limits to Growth*. New York: Universe Books, 1972.

Meadows, D.H., D.L. Meadows, and J. Randers. *Beyond the Limits*. Toronto: McClelland & Stewart Inc., 1992.

Mesarovic, M.D. *Mankind at the Turning Point*: first edition. New York: Dutton, 1974.

Michaels, P.J. "Crisis in Politics of Climate Change Looms on Horizon," in *Forum*, vol. 4, no. 4, Winter 1989: 14–23.

Michaels, P.J. *Sound & Fury*. Washington, DC: CATO Institute, 1992.

Miles, R. Jr. *Awakening from the American Dream*. New York: Universe Books, 1976.

Mills, C. "Why Life is Disappointing," in *Values & Public Policy*, C. Mills, ed. Orlando: Harcourt Brace Jovanovich Inc., 1992, 563–5.

Milne, A. "The Perils of Green Pessimism," in *New Scientist*, no. 138, 1993: 34–7.

North, R.D. *Life On a Modern Planet*. Manchester: Manchester University Press, 1995.

O'Riordan, T. and J. Cameron, eds. *Interpreting the Precautionary Principle*. Earthscan Publishers, 1994.

O'Riordan, T. and A. Jordan. "The Precautionary Principle in Contemporary Environmental Politics," in *Environmental Values*, vol. 4, no. 3, 1995: 191–212.

Oster, G.G. "Predicting Populations," in *American Zoologist*, no. 21, 1981: 831–44.

Pascal, B. *Pensées*, translated by A.J. Krailsheimer. London: Penguin Classics, 1966.

Peacock, K.A. "The Ozone Surprise," in *Living with the Earth*, K.A. Peacock, ed. Toronto: Harcourt Brace & Company, 1996.

Peacock, K.A., ed. *Living with the Earth*. Toronto: Harcourt Brace & Company, 1996.

Pestel, E. *Beyond the Limits to Growth*. New York: Universe Books, 1989.

Peterman, R.M. and M. M'Gonigle. "Statistical Power Analysis and the Precautionary Principle," in *Marine Pollution Bulletin*, vol. 24, no. 5, 1992, 231–4.

Peters, R.H. *A Critique For Ecology*. Cambridge: Cambridge University Press, 1991.

Picard, A. "Greenhouse Effect Blamed for Deaths of One Million in Third World Last Year," in *Living with the Earth,* K. A. Peacock, ed. Toronto: Harcourt Brace & Company, 1996, 35 (reprinted from *The Globe and Mail*, 24 November 1989).

Plato. *Apology* [translated by Hugh Tredennick], in *The Collected Dialogues of Plato*, E. Hamilton and H. Cairns, eds. Princeton, New Jersey: Princeton University Press, 1973, 29a-b, 3–26.

Popper, K. *The Open Universe*. London: Hutchinson, 1982.

Quine, W.V.O. "Two Dogmas of Empiricism," in *From a Logical Point of View*. Cambridge: Harvard University Press, 1953, 20–46.

Quine, W.V.O. and J.S. Ullian. *The Web of Belief*: second edition. New York: Random House, 1978.

Rescher, N. *Risk: A Philosophical Introduction to the Theory of Risk Evaluation and Management*. Washington, DC: University Press of America, 1983.

Roberts, P.C. *Modelling Large Systems*. London: Taylor and Francis, 1978.

Rothman, T. "A Memoir of Nuclear Winter," in *Science à la Mode*. Princeton, New Jersey: Princeton University Press, 1989, 109–47.

Sagoff, M. "The Limits of Cost-Benefit Analysis," in *Values & Public Policy*, C. Mills, ed. Orlando: Harcourt Brace Jovanovich Inc., 1992, 76–9.

Sagoff, M. "Fact and Value in Environmental Science," in *Environmental Ethics*, vol. 7, no. 2, 1995: 99–116.

Schell, J. "Our Fragile Earth," in *Discover Magazine*, vol. 10, no. 10, October 1989: 44–50.

Schneider, D. "The Rising Seas," in *Scientific American*, vol. 276, no. 3, March 1997: 112–17.

Schneider, S.H. "The Changing Climate," in *Scientific American*, vol. 261, no. 3, September 1989: 75.

Seller, A. "Realism versus Relativism: Towards a Politically Adequate Epistemology," in *Feminist Perspectives in Philosophy*, M. Griffiths and M. Whitford, eds. Indiana University Press, 1988: 169–86.

Shackley, S., P. Young, S. Parkinson, and B. Wynne. "Uncertainty, Complexity and Concepts of Good Science in Climate Change Modelling: Are GCMs the Best Tools?" in *Climatic Change*, no. 38, 1998: 159–205.

Shrader-Frechette, K.S. *Risk Analysis and Scientific Method.* Boston: D. Reidel Publishing Co., 1985.

Shrader-Frechette, K.S. *Risk and Rationality: Philosophical Foundations for Populist Reforms.* Berkeley: University of California Press, 1991.

Shrader-Frechette, K.S. and E.D. McCoy. *Method in Ecology.* Cambridge: Cambridge University Press, 1993.

Simon, J. *Population Matters.* New Jersey: Transaction Publishers, 1990.

Simon, J. and H. Kahn. *The Resourceful Earth.* Oxford: Blackwell, 1984.

Sjoberg, L., ed. *Risk and Society.* London: Allen & Unwin Ltd., 1987.

Stebbing, A.R.D. "Environmental Capacity and the Precautionary Principle," in *Marine Pollution Bulletin*, vol. 24, no. 6, June 1992, 287–94.

Swift, J. *Gulliver's Travels.* London: Penguin Books, 1967.

Thompson, P. *The Structure of Biological Theories.* Albany: State of New York Press, 1989.

Thompson, S.L. and S.H. Schneider. "Nuclear Winter Reappraised," in *Foreign Affairs*, vol. 64, no. 5, Summer 1986: 981–1005.

Thoreau, H.D. *Walden.* New York: Harper & Row Publishers, 1965.

Turco, R., O.B. Toon, T. Ackerman, J. Pollack, and C. Sagan. "Nuclear Winter: Global Consequences of Multiple Nuclear Weapons Explosions," in *Science*, 23 December 1983: 1283–92.

van Kooten, G.C. "Effective Economic Mechanisms: Efficiency and Ethical Considerations," in *Ethics & Climate Change*, H. Coward and T. Hurka, eds. Waterloo, Ontario: Wilfrid Laurier University Press, 1993, 133–48.

VanderZwaag. D. *CEPA and the Precautionary Principle Approach*, The Issues no.18, Government of Canada Publications, 1994.

Vanier, J. *In Weakness, Strength: The Spiritual Sources of Georges P. Vanier.* Toronto: Griffin House Press Ltd., 1969.

von Furstenberg, G.M. *Acting Under Uncertainty.* Kluwer Academic Publishers, 1990.

Watson, R.T., H. Rodhe, H. Oeschger, and U. Siegenthaler. "Greenhouse Gases and Aerosols." in *Climate Change: The IPCC Scientific Assessment*, Houghton et al., eds. Cambridge: Cambridge University Press, 1990, 1–40.

Wernham, J.C.S. *James's Will-to-Believe Doctrine: A Heretical View.* Kingston and Montreal: McGill-Queen's University Press, 1987.

Weyant, J.P. "The Costs of Carbon Emissions Reductions," in *Economics and Policy Issues in Climate Change*, W.D. Nordhaus ed. Washington, DC: Resources for the Future, 1998, 191–214.

Worzel, R. "The Sins of Eco-phonies," in *The Globe and Mail*, 27 November 1992: A26.

Index